William Henry Broadbent, John Francis Harpin Broadbent

Heart disease, with special reference to prognosis and treatment

William Henry Broadbent, John Francis Harpin Broadbent

Heart disease, with special reference to prognosis and treatment

ISBN/EAN: 9783744737173

Printed in Europe, USA, Canada, Australia, Japan

Cover: Foto ©berggeist007 / pixelio.de

More available books at **www.hansebooks.com**

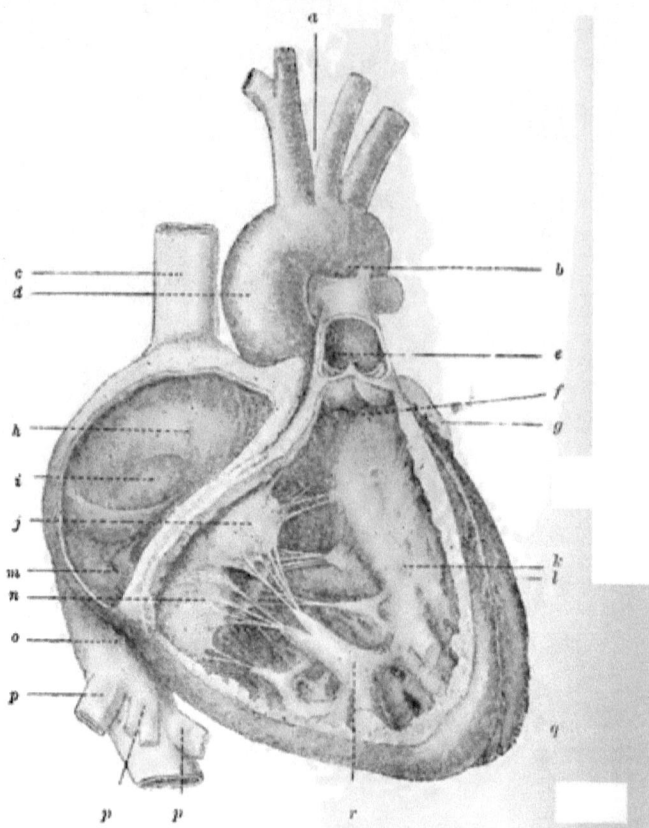

Description of Diagram.—The heart and great vessels, with the anterior walls of the right auricle and right ventricle dissected off to show position of the valves and septa. (Quain's *Anatomy.*)

a, Innominate and left carotid arteries; *b*, Transverse part of arch; *c*, Vena cava superior; *d*, Ascending part of arch of aorta; *e*, Pulmonary artery; *f*, Pulmonic valves; *g*, Appendix of left auricle; *h*, Inter-auricular septum; *i*, Fossa ovalis, with Eustachian valve below; *j*, Left segment of tricuspid valve; *k*, Inter-ventricular septum; *l*, Left ventricle; *m*, Coronary vein; *n*, Right segment of tricuspid valve; *o*, Inferior vena cava; *p*, Hepatic veins; *q*, Left ventricle; *r*, Anterior papillary muscle.

HEART DISEASE:

WITH SPECIAL REFERENCE TO

PROGNOSIS AND TREATMENT.

BY

SIR WILLIAM H. BROADBENT, Bart.

M.D. Lond., F.R.S., F.R.C.P.,

PHYSICIAN IN ORDINARY TO H.R.H. THE PRINCE OF WALES;
CONSULTING PHYSICIAN TO ST. MARY'S HOSPITAL AND THE LONDON FEVER HOSPITAL;
LATE PRESIDENT OF CLINICAL, MEDICAL, NEUROLOGICAL AND HARVEIAN SOCIETIES;

AND

JOHN F. H. BROADBENT,

M.A., M.D., (Oxon.), M.R.C.P.

NEW YORK
WILLIAM WOOD & COMPANY
1897

PREFACE.

This book is, for the most part, a reproduction of lectures on "Prognosis in Valvular Disease of the Heart," delivered before the Harveian Society in 1884, and of the Lumleian Lectures at the Royal College of Physicians on "Prognosis in Structural Diseases of the Heart," delivered in 1891.

The prognosis of heart disease already engaged my attention when I was house-physician under Sibson at St. Mary's Hospital, and my first paper on this subject was read before the Harveian Society in 1866. Up to that time there had not, so far as I am aware, been any systematic study and exposition of the indications by which the probable course of disease of the heart in different cases might be foreseen, and ideas which tended to obscure the interpretation of the symptoms and physical signs were held by physicians of great experience and authority. Traces of the controversies of those days will still be found in the present work. They have lost much of their interest, but references to them could not well be entirely omitted.

The prognosis of heart disease is worthy of special study, not only on account of its inherent importance, but also because the knowledge which enables the medical man to forecast clearly the course of the disease constitutes the best preparation for its treatment. The subject of treatment did not, however, enter into the scheme of the lectures, but it is engrafted upon them in the present work, and for this, and for the rearrangement rendered necessary by it, I am indebted to my son, Dr. John Broadbent, without whose efficient assistance and co-operation the task of preparing this book could not have been accomplished.

The lectures on which the book is based having been addressed to the College of Physicians and to the Harveian Society, presupposed a knowledge of heart disease on the part of my audience, and a minute exposition and analysis of the symptoms and physical signs, by means of which the diagnosis of the different valvular and structural affections of the heart is arrived at, were therefore unnecessary. This is still taken for granted, but a brief chapter on the examination of the cardiac region, and on the significance of the various departures from normal conditions, has been added.

The principal motive for reproducing the lectures has been the frequently expressed wish on the part of old pupils to see my teachings on heart disease in a collected form. They have had to wait long, and to them I now dedicate the book.

W. H. BROADBENT.

Sept. 15th, 1897.

CONTENTS.

CHAPTER I.

The relations of the heart to the chest walls—On the methods of Examination by inspection: palpation, percussion, and auscultation 11

CHAPTER II.

VALVULAR DISEASE.

Relative frequency of the different lesions—Points to be considered in studying a case of valvular disease—The physical signs—Cardiac murmurs, their significance as regards the seat and as regards the extent of the lesion—Modification of heart-sounds—The pulse, its importance in diagnosis 26

CHAPTER III.

Hypertrophy and dilatation of the heart: their importance as a means of estimating the extent of a valvular lesion—Possible objections to this view discussed—Explanation of the way in which the different valvular lesions give rise to hypertrophy and dilatation: (1) Aortic stenosis; (2) Aortic incompetence; (3) Mitral regurgitation; (4) Mitral obstruction—Compensation 38

CHAPTER IV.

Symptoms in aortic and in mitral disease—Dropsy: discussion as to its causation 48

CHAPTER V.

Ætiology of valvular lesions—Acute endocarditis—Chronic endocarditis: degenerative changes—Rupture of valve—Dilatation of the orifice, secondary to cardiac dilatation—Congenital lesions of valves 60

CHAPTER VI.

Prognosis in valvular disease—The nature of the lesion: the relative danger attaching to each particular lesion—Sudden death: the valvular diseases in which it is liable to occur—The extent of the lesion—The stationary or progressive character of the lesion as influencing prognosis 67

CHAPTER VII.

Prognosis continued—Age, sex, heredity—Effects of high arterial tension—Habits and mode of life of the patient—Anæmia—The circumstances under which prognosis may have to be made: (1) Immediately after acute endocarditis; (2) When the valvular lesion is slight and has given rise to no structural changes in the heart; (3) When compensatory changes have taken place but no symptoms of embarrassment of the circulation are present; (4) When symptoms of failure of compensation have set in; (5) In advanced valvular disease when severe symptoms of cardiac failure have supervened 78

CHAPTER VIII.

TREATMENT.

Treatment of valvular disease in general—Prophylactic measures—General rules, in cases where lesion is not of serious extent, as to exercise: Œrtel and Schott treatments, climate, choice of residence, diet, stimulants—Treatment of anæmia as complication—Treatment where lesion is of more serious nature and has given rise to marked hypertrophy and dilatation of the heart—Precautions to be taken—Selection of winter resort—Importance of rest—Diet—Stimulants—Employment of drugs—Purgatives, their importance—Treatment of venous congestion—Venæsection—Treatment of the condition of asystole in aortic disease—Digitalis in aortic incompetence 88

CHAPTER IX.

Abuse of digitalis—Substitutes for digitalis—The group of cardiac tonics of the digitalis type—their physiological action: therapeutic effects—Use of digitalis in aortic stenosis, in mitral incompetence, in mitral stenosis 112

CHAPTER X.

THE INDIVIDUAL VALVULAR LESIONS.—AORTIC STENOSIS.

The murmur of aortic stenosis—Conditions other than aortic stenosis which may give rise to systolic aortic murmurs—Causes of aortic obstruction—Differential diagnosis of murmurs—Estimation of extent of lesion by means of the murmur, the changes in the heart, the pulse—Progress of the disease: Symptoms—Prognosis—Treatment 124

CHAPTER XI.

AORTIC INCOMPETENCE.

Physical signs—The diastolic murmur: directions in which it is conducted—Presystolic murmur, its significance—Modification of the aortic second sound—Pulsation of arteries—Capillary pulsation—The collapsing pulse—Pulsus bisferiens—Irregular pulse—Estimation of the amount of regurgitation by the character of the murmur, of the aortic second sound, of the pulse, and by the changes in the heart—Aortic incompetence due to syphilis and causes other than acute endocarditis—Symptoms—Prognosis—Treatment 137

CHAPTER XII.

MITRAL INCOMPETENCE OR REGURGITATION.

Physical signs—The murmur of mitral incompetence—The pulse—Explanation of irregularity of pulse—Mitral incompetence due to actual change in valves, the result of endocarditis—Estimation of extent of lesion: (1) From character of murmur and first sound; (2) From compensatory changes in the heart; (3) From symptoms—Prognosis—Mitral incompetence without damage to valves—Its causation and explanation—Differential diagnosis—The mitral incompetence of middle or old age—Treatment 162

CHAPTER XIII.

MITRAL STENOSIS.

Its predominance in the female sex—Morbid anatomy and physiology of constriction of the mitral orifice—The physical signs—The pulse—The changes in the heart—The cardiac murmurs—Three stages in the progress of the disease as defined by auscultatory signs—The characteristics of these three stages—Symptoms—Diagnosis—Prognosis—Treatment 185

CHAPTER XIV.

VALVULAR DISEASE OF THE RIGHT SIDE OF THE HEART.

Tricuspid incompetence and stenosis—Pulmonic incompetence and stenosis—Systolic pulmonic murmurs which do not indicate stenosis 207

CHAPTER XV.

CONGENITAL MALFORMATIONS.

Varieties of congenital malformations—Relative frequency of occurrence—Of single and combined defects—Cyanosis—Cause of Cyanosis—Physical signs—Symptoms—Diagnosis—Prognosis 213

CHAPTER XVI.
ADHERENT PERICARDIUM.

Morbid anatomy—Physical signs—Symptoms—Diagnosis—Prognosis—Treatment ... 220

CHAPTER XVII.
STRUCTURAL DISEASE OF THE HEART.

Hypertrophy—Causes of hypertrophy of the left ventricle—Physical signs—Symptoms—Prognosis—Treatment—Dilatation—Causes of dilatation—Illustrative cases—Physical signs—Symptoms—Prognosis—Treatment ... 226

CHAPTER XVIII.
STRUCTURAL DISEASE OF THE RIGHT VENTRICLE.

Physiological dilatation—Hypertrophy and dilatation—Degeneration of the muscular walls ... 272

CHAPTER XIX.
FATTY DEGENERATION.

Distinction between fatty infiltration of obesity and fatty degeneration—Causation of fatty degeneration—Symptoms—Physical signs—Diagnosis—Prognosis—Treatment ... 282

CHAPTER XX.
ANGINA PECTORIS.

Characteristics of true anginoid pain—Duration of attack—Aspect of patient—Exciting causes of paroxysm—Pathology and ætiology of true angina—Theories as to cause of the pain—Prognosis—Treatment ... 296

CHAPTER XXI.
FUNCTIONAL AFFECTIONS (SO-CALLED) OF THE HEART.

Pain in præcordial region—Its causes and treatment—Palpitation: (1) Persistent tachycardia; (2) temporary intermittent attacks of palpitation—Causation—Treatment—Intermittent and irregular action of the heart ... 315

APPENDIX.

Note on the preparation of the Baths, and on the movements practised in the Schott treatment of heart disease ... 325

INDEX ... 329

HEART DISEASE.

CHAPTER I.

THE RELATIONS OF THE HEART TO THE CHEST WALLS—
ON THE METHODS OF EXAMINATION BY INSPECTION;
PALPATION, PERCUSSION, AND AUSCULTATION.

A SYSTEMATIC description of the anatomy and relations of the heart, and a text-book exposition of the physical signs by means of which departures from the normal condition of the heart are recognized, would be out of place in a work of this character. A brief reminder, however, of the relations of the heart to the chest wall, together with a short description of the method in which a clinical examination of the heart should be conducted, and an indication of the significance of the various points observed, may form a useful preliminary to the study of heart disease. This will therefore form the subject of the introductory chapter.

The first portion of this chapter, dealing with the position and relations of the heart, is mainly taken from the works of Sibson, who was a most careful and accurate observer, and devoted much time and labour to the study of this subject.*

THE POSITION OF THE HEART AND THE RELATION OF ITS CAVITIES TO THE CHEST WALLS.

The heart and great vessels, with their pericardial covering, occupy the central portion of the thoracic cavity.

* Sibson's Works, edited by Ord. Vol. iii.

The right auricle and ventricle compose the whole of the front of the heart, with the exception of a narrow strip to the left, where the left ventricle comes into view from behind.

The **right auricle** lies behind the sternum, its upper border being on a level with the third costal cartilages, and extending from a point about one inch to the right of the sternum nearly to its left border. It is broadest above, and narrows down almost to a point at the lower end of the auriculo-ventricular furrow, which forms its left border and runs obliquely downwards from the sternal end of the third left to the sternal end of the seventh right costal cartilage. Its right border is convex, and lies behind the right costal cartilages, just beyond the right margin of the sternum.

The right auricle undergoes more change in form during the action of the heart than any other portion of the organ, becoming nearly twice as wide in systole of the ventricles as in diastole. The auriculo-ventricular furrow also sweeps backwards and forwards to so great an extent, to the left during systole and to the right during diastole of the ventricles, that it presents no fixed position during life.

The **right ventricle**, when exposed to view, presents a pyramidal shape. The base of the pyramid is formed by the lower boundary of the ventricle, which rests on the central tendon of the diaphragm; the apex of the pyramid is crowned and completed by the pulmonary artery; the left side is formed by the furrow which divides the left from the right ventricle; the right side, by the furrow separating the right auricle and ventricle.

The furrow separating the two ventricles runs from the third to the fifth left costal cartilages, just behind their junction with the ribs.

The **left ventricle** forms the convex left border of the heart as seen from the front, and forms a long, narrow strip extending from the third left intercostal space down to the fifth, where it terminates in the apex of the heart. This occupies the fifth space, being usually situated just internal to a line drawn vertically downwards from the nipple.

The **appendix of the left auricle** lies just behind the third left costal cartilage, close to its junction with the third rib, and fills up the space between the upper end of the left and right ventricles at the top of the longitudinal or interventricular furrow.

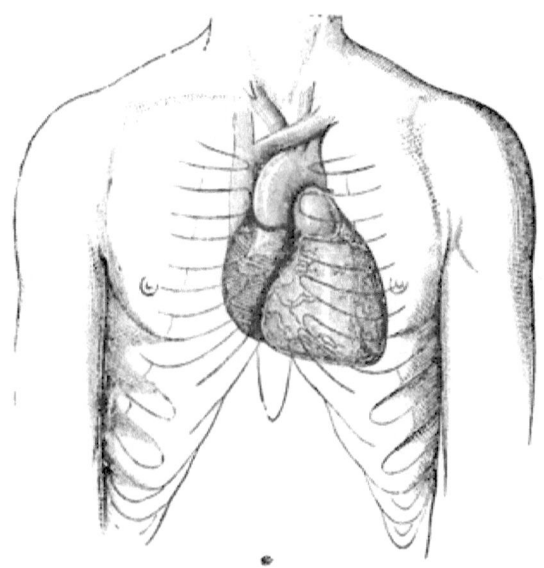

FIG. 1.—RELATION OF HEART TO CHEST WALLS (*Sibson*).

The area which would be marked out on the chest wall by percussion, indicating the position of the heart beneath and its relation to the chest wall, is termed "the area of deep cardiac dulness."

The lungs cover the great vessels and the whole of the heart except a portion of the right ventricle.

The inner margins of the right and left lungs in front meet behind the sternum for its upper two-thirds. The inner margin of the left lung diverges from that of the right at the level of the fourth left costal cartilage. It passes thence to the left along the lower edge of the fourth cartilage in front of the body of the right ventricle, curving downwards just before it reaches the junction of the cartilage with the rib; it then crosses the fourth space and

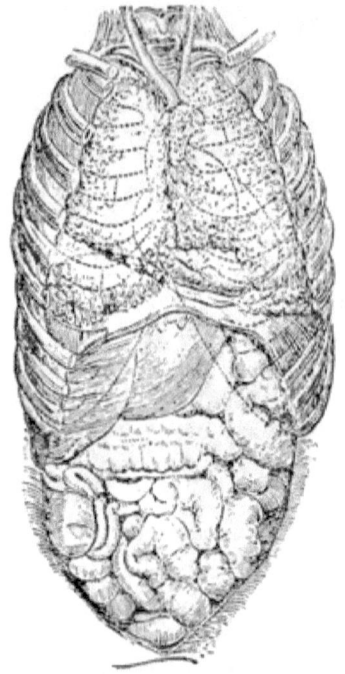 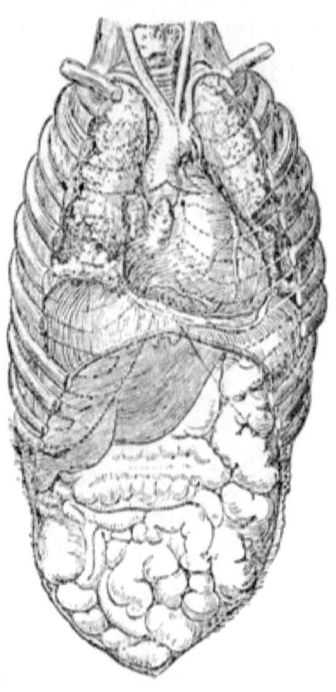

FIG. 2.—THE HEART WITH LUNGS IN SITU TO ILLUSTRATE AREA OF SUPERFICIAL CARDIAC DULNESS (*Sibson*).

FIG. 3.—THE HEART WITH OVERLYING LUNG DISSECTED OFF TO ILLUSTRATE AREA OF DEEP DULNESS (*Sibson*).

fifth cartilage, and curving, so as to form a hollow space for the lodgment of the apex, ends behind the sixth cartilage.

The inner margin of the right lung continues its course nearly straight downwards behind and a little to the left of the centre of the sternum, to the level of the sixth chondrosternal articulation. It thus covers the right auricle, the auriculo-ventricular furrow, and the right border of the right ventricle.

The area on the chest wall corresponding to the part of the heart uncovered by lung can be accurately marked out by percussion, and is known as "the area of superficial cardiac dulness."

The area thus indicated would be mapped out as follows: its right margin, by a line running from above downwards slightly to the left of the middle line of the sternum from the level of the fourth costal cartilage down to that of the sixth; its highest point would be at the level of the fourth costal cartilage; its left limit would be defined by a line running outwards from the sternum at this level, along the lower edge of the fourth left costal cartilage, nearly as far as its junction with the rib, and then dipping downwards and curving slightly outwards to the point on the chest wall where the apex beat is felt. The base or lower limit corresponds to a line drawn from the sixth right costal cartilage to the apex.

The Clinical Examination of the Heart.

This is conducted by inspection, palpation, percussion, and auscultation. Much is learnt by careful inspection. Bulging or other general feature of the præcordial region is to be noted. Any deformity of the chest giving rise to displacement of the heart, or fixation of one side of the thorax by pleural adhesions, or by fibroid or other disease of the lung, must be taken into account. The position and character of the apex beat should be determined as far as this is possible; the surface of the chest should be carefully scrutinized for pulsation in abnormal situations, and epigastric pulsation or any visible heave over the right ventricle should be noted.

These observations will be corroborated or checked by palpation; but there are other points which can be ascertained by the eye alone, such as retraction of intercostal spaces. This is most commonly seen in the third, fourth, and fifth left intercostal spaces and is usually systolic in time; it may either arise from a direct tug on the spaces by pericardial adhesions, or be due to atmospheric pressure: the latter condition is more common, and in such cases the heart will usually be dilated and hypertrophied, and therefore subject to great diminution in volume during systole.

Inspection also embraces the neck, where is seen the carotid throb which betrays aortic regurgitation. To be characteristic, it must be visible, not at the root of the neck only, but as high as the hollow between the ramus of the jaw and the sternomastoid. It may, exceptionally, be present where there is no aortic disease, and it may be simulated by pulsation in the deep jugular vein. This, however, is easily distinguished, being readily arrested by light pressure.

Important information may frequently be derived from careful observation of the veins of the neck. The jugulars on one or both sides may be full and distended. If the distension is present on one side only, it may be due to pressure on the innominate vein of that side by an aneurysm or some other intra-thoracic tumour. If it exists on both sides, it may be due either to pressure on the superior vena cava or both innominate veins, or to dilatation and engorgement of the right side of the heart from back pressure through the lungs. In the latter case, the veins will be temporarily more or less emptied by a forcible deep inspiration, and will become more distended during expiration: in the former, variations of the intra-thoracic pressure during inspiration and expiration will have no appreciable effect in reducing or augmenting the distension. Pulsation of the jugular veins may be present when the right side of the heart is greatly dilated and tricuspid regurgitation has set in. The pulsation is often double, both auricular and ventricular systoles transmitting backward waves. It has been studied with great care by Mackenzie, of Burnley, who has obtained graphic records of two distinct types of jugular pulsation, and established their significance.

Finally, mention may be made of enlarged and tortuous veins on the surface of the chest, constituting collateral channels between the superior and inferior venæ cavæ when one or other is compressed, and of pulsating veins occasionally seen dipping into an intercostal space near the sternum.

PALPATION.

The information obtained by palpation is second in importance only to that afforded by auscultation. The situation, force, limits, and extent of the cardiac impulse are accurately ascertained, and the relative intensity of the apex beat and right ventricle beat must be compared. The examination, to be complete, must be conducted in various ways.

Firstly, the flat of the right hand should be placed over the cardiac area, the fingers covering the apex region. The powerful heaving impulse of the apex in hypertrophy, the diffuse shock or slap in dilatation, and the peculiar sudden tap in mitral stenosis are often at once diagnostic: a powerful heave of the lower left costal cartilages, and sometimes of the lower end of the sternum as well, will indicate stress thrown on the right ventricle by back pressure through the pulmonary circulation. Thrills or vibrations, systolic or presystolic, over the area of the apex beat or elsewhere will be recognized at the same time. Sometimes a peculiar diastolic shock or retraction may be felt over the apex or the right ventricle, caused by the presence of pericardial adhesions. Next the fingers must be pressed into the right intercostal spaces in search of pulsation, and the apex region must be explored with the tips of the fingers, and the exact position, extent, force, and deliberate or sudden character of the apex beat must be noted. This investigation must not be limited to the point at which the apex appeared to present itself, when the hand was applied to the præcordium; the real apex beat may be sometimes far out in the axilla, or high up in the fourth space, or it may be concealed behind the mamma.

Speaking generally, displacement of the apex downwards is indicative of hypertrophy of the left ventricle, dilatation giving rise to extension of the impulse outwards and downwards; displacement of the apex to the left is caused by enlargement of the right ventricle, which is usually due to combined hypertrophy and dilatation. For the most part, any considerable change in the position of the apex beat is the result of hypertrophy and dilatation of both sides of the heart. Special displacement of the apex not due to intrinsic changes in the heart may be the result of various causes, such as pericardial or pleuro-pericardial adhesions, pleural effusion, pneumo-thorax, fibroid condition or cavitation of the lung, or deformity of the chest. It is always well to ascertain whether the apex changes its position in respiration, and with change of posture of the patient from the erect to the recumbent position, and when lying first on one side, then on the other.

It is not uncommon to find that neither apex beat nor impulse of any kind is to be felt. This does not necessarily imply feeble action of the heart, but may be due to thickness of the parietes or great depth of thorax, or to overlapping emphysematous lung, or to pericardial adhesions.

Departures from the normal in the cardiac rhythm should also be noted; they are of two kinds: Irregularity and Intermission—irregularity, when the beats are of varying force and follow each other at unequal intervals; intermission, when a beat is dropped from time to time, the intervening beats being of equal force and at equal intervals.

The irregular action of the heart is usually exaggerated in the pulse, as many of the beats may fail to reach the wrist. When the pulse is intermittent, it will usually be found that there is not an entire absence of the beat of the heart, but an imperfect and hurried beat corresponding to the intermission.

The examination by palpation, to be complete, must embrace the abdomen, and the size of the liver must be carefully ascertained. As the right ventricle begins to fail, and obstruction to the return of blood to the heart from the liver through the inferior vena cava begins to be felt, the liver becomes engorged and congested, and gradually increases in size, till it may become very much enlarged, extending even below the umbilicus. The size of the liver is therefore an important index as to the degree of venous obstruction.

In examining to ascertain the size of the liver, the patient should be made to lie on his back with the legs drawn up, so that the belly is perfectly flaccid. The flat of the hand should then be placed on the abdomen, and the edge of the liver carefully felt for; at the same time the other hand should be placed beneath the back of the patient below the false ribs, and the liver lifted from behind, so that if the liver edge is not at first felt, the jogging from behind will press it up against the hand on the abdomen, and render it more obvious. Palpation with the tips of the fingers will probably provoke some rigidity of the abdominal walls, which obscures the resistance presented by the liver behind. When, however, there is fluid present in the abdominal cavity, this method of palpation will be useful, for then on dipping sharply down with the fingers, the fluid between the liver and abdominal walls will be displaced, and the firm, hard substance of the liver be felt below.

Percussion is, as a rule, a very untrustworthy method of ascertaining the size of the liver, owing to conducted resonance from distended coils of intestine; it will, however, serve to mark out the upper margin of the liver, unless there be fluid or old pleural adhesions at the base of the right pleural cavity.

Pulsation of liver is best detected by combined simultaneous palpation and inspection. The hand should be placed on a part of the liver at some distance from the epigastrium, and gentle pressure being made, it should be carefully watched. If the liver is actually pulsating, and not merely jogged by a hypertrophied right ventricle, the hand will be seen to rise and fall rhythmically even though no actual pulsation is felt on palpation.

Percussion.

Percussion will verify and perhaps extend the information obtained by inspection and palpation. It is specially useful when no impulse is present. There is normally, as has already been described, a small area of absolute dulness—"the superficial cardiac dulness"—where the heart is not covered by lung, which is easily defined. It may be encroached upon and obliterated when the lungs are emphysematous, or extended when they are shrunken as a result of fibroid contraction or pleural adhesion. The deep dulness indicating the true dimensions of the heart cannot be mapped out with absolute accuracy in all cases, as the overlying lung tissue varies in thickness and extent in different individuals, and in inspiration and expiration. In percussing for this purpose, the finger must be pressed firmly into the intercostal spaces, and the stroke must be delivered smartly and perpendicularly. The note changes gradually as the dull area is left, with the increase in the thickness of the overlapping lung, till full lung resonance is reached. The end of expiration, when the cushion of lung is thinnest, should be chosen as the moment to attempt the definition of the comparative dulness which indicates the size and position of the heart. The so-called auscultatory percussion is, in my opinion, of no value whatever.

AUSCULTATION.

The physical examination is completed by auscultation. For the most part, a diagnosis has already been arrived at before the stethoscope is applied, and in all cases the information obtained by auscultation must be checked and interpreted by evidence derived from the pulse, and from inspection, palpation, and percussion. Either a flexible or rigid stethoscope may be employed. The former is more convenient and expeditious, as it can be shifted from point to point without moving the head, and it is always under the eye, so that bulging or retraction of spaces can be timed and co-ordinated with the sounds or murmurs heard. On the other hand, it is sometimes difficult to distinguish between the first and second sound, and the rigid stethoscope, by communicating a faint jar to the ear, at once points out the sound which is produced by the systole.

The stethoscope should be applied successively to the apex region, the tricuspid, pulmonic, and aortic areas, and the character, and absolute and relative loudness of the sounds should be noted at each point; observations taken simply at the apex or base afford very imperfect information.

THE MITRAL AREA.

At the apex both sounds are normally audible—the first, comparatively long and low-pitched; the second, short and sharp. The first sound may undergo various modifications. It may be prolonged, which is usually indicative of hypertrophy; or, on the other hand, it may be short and sharp. If when short it is also loud, it is among the indications of dilatation; if weak, it may be due either to degeneration or to simple asthenia of the cardiac muscle. A peculiar sharp and snapping character of the first sound at the apex is produced by mitral stenosis. Reduplication of the first sound is an important modification often met with; it is most distinct when the stethoscope is placed exactly over the septum between the apex and right ventricle, i.e. partly over one ventricle, partly over the other. Left of this spot the sound may be merely blurred; right of it, reduplication may still be distinct. The double sound is due to want of synchronism between the two ventricles, and may be produced whenever undue stress is put upon the left ventricle by high systemic arterial tension, or on the right by obstruction in the pulmonary circulation. The second sound heard at the apex is, for the most part, that produced at the aortic valves; occasionally it is more distinct here than at the aortic area proper. Its modifications will be described when the sounds at the base are discussed.

The interval between the first and second sounds indicating the duration of the cardiac systole, and that between the second and first sounds, or the diastolic interval, should be carefully observed, and any deviation from the normal noted. The rhythm may be disturbed in one direction by the shortening of the diastolic interval, till the sounds are equidistant and resemble the ticking of a watch, as in palpitation, or the sounds may become equidistant from prolongation of the systole, when they resemble the beat of a pendulum. In the other direction the systolic interval may be shortened till the second sound follows the first almost without an appreciable pause: this is a serious condition, indicating incomplete contraction of the ventricle, and also, in many cases, impending cardiac failure.

The murmurs heard at the apex may be systolic, presystolic, or diastolic; of these the most frequent is the systolic. It is most commonly smooth and blowing in character; sometimes a musical element may be present, or the murmur may be rumbling and indistinct; it is rarely rough and croaking, as is often the case with aortic systolic murmurs. The systolic murmur heard at the apex always signifies more or less regurgitation through the mitral orifice, and it will be necessary to note how far outside the apex the murmur is audible, and to what extent it replaces the first sound, as these will be important points in estimating its significance. An apex murmur is sometimes more distinct in the recumbent than in the sitting or erect position, or may be brought out by slight exertion.

A pulmonic or aortic systolic murmur may be conducted to the apex, but it will be of diminished intensity and will not, as a rule, be conducted towards the axilla.

The presystolic murmur, so named from the fact that it precedes and runs up to the first sound produced by the cardiac systole, is usually vibratory in character. Speaking generally, it indicates the presence of mitral stenosis, and is produced by the rush of blood from the auricle into the ventricle through the narrowed mitral orifice; but it may be present as a temporary phenomenon after acute disease. In an early stage it is brief in duration, and rises rapidly in intensity, terminating abruptly in the first sound, when it corresponds with and is produced by the auricular systole. It may, however, be greatly prolonged; that is, it may begin at a much earlier period so as to correspond with the active dilatation stage of the ventricular diastole, which precedes the auricular systole.

A diastolic murmur audible at the apex is usually that of aortic insufficiency, conducted from the base to the apex by the walls of the heart.

The Tricuspid Area.

The sounds heard over the tricuspid area are mainly those of the right ventricle; the first sound is rather shorter and louder than that at the apex, and the second sound is intensified when there is obstruction in the pulmonary circulation.

Systolic, presystolic, and diastolic murmurs may be heard in this region, and the former may indicate tricuspid incompetence, but the two latter varieties are most commonly murmurs conducted from the apex or base, and cannot be relied upon for diagnostic purposes.

The Aortic Area.

In the pulmonic and aortic areas accentuation of the second sound is the point specially to be observed, denoting increased pressure in the pulmonic or systemic circulation respectively. Reduplication of the second sound is not uncommon, and is due to a synchronous closure of the aortic and pulmonic semilunar valves.

The murmurs audible over the aortic area in the second right intercostal space may be systolic or diastolic. A systolic murmur may indicate actual obstruction, but is more frequently due to mere roughening or rigidity of the cusps of the valves, or to dilatation of the aorta beyond the valves, or other conditions not necessarily causing narrowing of the orifice. It may be loud and rough, or musical, or soft blowing. A diastolic murmur is usually blowing in character, and almost invariably indicates aortic incompetence, being produced by the regurgitant stream of blood; it can sometimes be heard lower down over the sternum, or a little to one side of it, at a point nearer the orifice of the aorta, when it is not audible in the so-called aortic area.

The Pulmonic Area.

Systolic and diastolic murmurs may also be heard in the pulmonic area, the third left intercostal space, and may indicate stenosis or incompetence of the pulmonic valves. These affections are, however, rare, and are usually of congenital origin. The murmurs heard in the pulmonic region are for the most part either conducted from the region of the aorta or apex, or are hæmic murmurs dependent on abnormal conditions of the blood, or they may occasionally be due to vibrations caused by the contact of the conus arteriosus with the chest wall. A diastolic aortic murmur is frequently louder in the pulmonic area than at the right second space, and may be audible here when it cannot be heard in the aortic area.

CHAPTER II.

VALVULAR DISEASE.

RELATIVE FREQUENCY OF THE DIFFERENT LESIONS — POINTS TO BE CONSIDERED IN STUDYING A CASE OF VALVULAR DISEASE—THE PHYSICAL SIGNS—CARDIAC MURMURS, THEIR SIGNIFICANCE AS REGARDS THE SEAT AND AS REGARDS THE EXTENT OF THE LESION — MODIFICATION OF HEART-SOUNDS — THE PULSE, ITS IMPORTANCE IN DIAGNOSIS.

DISEASES of the heart are classified as structural and valvular, according as the morbid change affects the muscular walls or the valves. There are, again, functional derangements which must be included among the affections of the heart.

The valvular lesions will be first discussed: they are what is usually understood by heart disease, and are most important because most numerous. We know more about them, and can be more certain of their diagnosis, thanks to the murmurs to which alterations in the valves give rise. We are also in a better position to estimate the extent of the lesion, and to give an accurate prognosis in valvular disease than in structural. The estimation of the obstruction produced by a certain degree of narrowing of one or other orifice, or of the amount of reflux resulting from incompetence of a given valve, is a problem of hydrostatics capable of solution, but no such definite conclusion can be arrived at, when in structural disease we have to form an opinion as to the contractile power and durability of muscular fibres in a certain stage of degeneration.

The Relative Frequency of the several Valvular Lesions.

Lesions of the valves of the left side of the heart constitute by far the most numerous and most important class of valvular affections. Primary valvular diseases of the right ventricle are comparatively rare; but tricuspid incompetence occurs not infrequently as a secondary result of antecedent valvular disease of the left side of the heart or of pulmonary disease.

The order of frequency, according to Walshe, is mitral regurgitation, aortic constriction, aortic regurgitation, mitral constriction, tricuspid regurgitation, pulmonic constriction, pulmonic regurgitation, tricuspid constriction.

The more frequent combinations of two or more are, according to the same authority, aortic constriction and mitral regurgitation; aortic constriction and regurgitation, mitral regurgitation and aortic regurgitation; mitral regurgitation and aortic constriction and regurgitation; mitral regurgitation and constriction with aortic constriction and tricuspid regurgitation; tricuspid and mitral constriction.

With regard to this classification, the first thing to be remarked is that aortic stenosis should certainly not stand in the second place in the order of frequency of valvular lesions: according to my experience, it is comparatively rare to find a case of pure aortic constriction. The explanation of the position assigned to it by Walshe is, doubtless, that he classed all cases in which a left basic systolic murmur was audible, as cases of aortic stenosis. But this murmur is producible by other causes than actual obstruction. I have long recognized, from bedside observation, as well as in the post-mortem room, and have taught for many years, that, while a systolic aortic murmur is one of the most common of the valvular murmurs, actual narrowing of the orifice is among the most rare of the valvular changes, and I should therefore place it third, or even fourth, in the list.

Some post-mortem statistics bearing on this question, compiled for me by Dr. Sidney Phillips a few years ago, may be interesting. Of 151 cases of morbus cordis, 11 only were cases of aortic stenosis, 49 were cases of mitral regurgitation, 53 cases of mitral stenosis, 38 cases of aortic regurgitation. These numbers are not given as representing with accuracy the proportionate frequency of valvular diseases, but they afford confirmation of the statement that aortic stenosis is the least common variety. Aortic constriction, therefore, whether singly or in combination with another lesion, should be placed much lower down on the list. According to my experience the following would be their order, mitral regurgitation, mitral constriction, aortic regurgitation, aortic stenosis. Of the combination of two or more lesions, mitral incompetence and stenosis, aortic and mitral regurgitation, aortic incompetence and stenosis, would occupy the first three places. By far the commonest lesion of the valves of the right ventricle is tricuspid incompetence, not occurring as a primary lesion, but as secondary, either to severe valvular disease of the left ventricle, or to lung disease, such as chronic bronchitis, which gives rise to obstruction to the flow of blood through the pulmonary circulation. Other valvular lesions of the right ventricle are comparatively rare, but I should place tricuspid stenosis, which may be associated with mitral stenosis, before pulmonic constriction or regurgitation.

The Clinical Study of Valvular Disease.

In studying a case of valvular disease of the heart, the following are the points which must be taken into consideration:—

    1. The valve affected, and the relative danger attaching to the particular lesion.

2. The actual condition of the orifice and valve—the degree of obstruction or amount of regurgitation to which the lesion has given rise.
3. The origin of the lesion, whether due to acute rheumatism, degenerative changes, or other causes.
4. The degree of soundness and vigour, functional and nutritional; firstly, of the muscular substance of the heart itself; secondly, of the tissues generally. How far, in fact, and for how long compensatory changes can be counted upon. In considering this question, family history will have an important place.

LOCALIZATION OF THE VALVULAR DISEASE.—SIGNIFICANCE OF THE CARDIAC MURMURS.

The chief guide in localizing disease in the valves of the heart is a murmur, produced, either by obstruction to the current of blood when one or other orifice is narrowed or roughened, or by regurgitation of the blood when a valve no longer closes perfectly. The term stenosis or constriction is employed to denote the condition of an orifice which is narrowed, the result of the narrowing being obstruction, and the term insufficiency or incompetence is employed to characterize the state of a valve which fails to close the opening it ought to protect, while regurgitation or reflux expresses the functional effect. It is well, as far as possible, to observe the distinction between the names indicative of structural change and those expressive of functional derangement resulting therefrom, but the terms "obstruction" and "regurgitation" are in such familiar use that they are frequently employed when "stenosis" and "insufficiency" would be more exact.

By means of the murmurs we learn definitely which valve is affected and what is the nature of the affection—whether such as to produce obstruction or regurgitation—

but they fail altogether by themselves to indicate the amount of damage which a valve has sustained. A loud murmur may be produced by a very slight change, and a murmur which is scarcely audible may be indicative of most extensive destruction of valves. Some information, however, may be gathered from a careful study of murmurs.

Different Characters of Murmurs: their Value as regards Estimation of the Extent of the Valvular Lesion.

Murmurs may be compared or contrasted in several respects: in intensity, they may be loud or soft; in duration, they may be long or short; they may be blowing, or musical, or mixed, rough and vibratory, or smooth. Again, they may begin with an accent, or rise gradually to a maximum intensity.

A loud murmur is, on the whole, of less serious import than one which is weak and soft. It is, at any rate, indicative of force in the heart's action, and of vigour in the movement of the blood; and weakness of the heart constitutes the greatest of all dangers. Then, again, although a rough edge to a large opening in aortic or mitral incompetence may generate vibrations which will produce a loud murmur, a mere slit in a membranous valve, or a shred of fibrin hanging by one end, is more likely to have such an effect.

In a case seen with Dr. Parrott of Hayes, a systolic mitral murmur, of such intensity as to be heard distinctly across a billiard-table, had been present for fifteen or twenty years without giving rise to symptoms, until dilatation of the heart was induced by extreme over-exertion. Another very loud murmur which came under my notice was attended with no important symptoms for the several months during which the patient was under observation. It was systolic aortic, and could be heard half a yard or more from the chest, through the man's clothes. In another case, a murmur almost as loud was found, after death, to be due to a delicate fibrous thread at the free margin of one of the aortic valves. The patient was sent into St. Mary's Hospital for heart disease and dropsy; but the dropsy was hepatic, and it was at once seen that the heart had no share in the production of the symptoms. The cause of death was cirrhosis of the liver, and there was no heart disease beyond the above-mentioned condition of one of the aortic valves. On the other hand, cases are met with in which the pulse may indicate serious aortic regurgitation while no diastolic murmur can be heard, and a murmur gradually develops itself as the patient gains strength and recovers from a state of extreme prostration; and it is common in mitral stenosis for murmurs to disappear with the supervention of serious symptoms, and to reappear as these are abated by treatment. It must not be concluded that a soft or weak murmur is necessarily indicative of either a failing heart or greatly damaged valve; but a diminution in the intensity of a murmur, gradual or sudden, may confirm unfavourable indications given by symptoms.

The **character of a murmur**—its roughness or vibratory character or smoothness—may have diagnostic significance, as will be pointed out later, but it does not give any information with regard to the extent of structural change or functional derangement. A musical murmur would seem to require for its production either a very fine chink with thin margins, or a thin membrane capable of vibrating like a string, and would therefore seem to be inconsistent with serious disease; but this cannot be laid down as an absolute rule. A musical note is often heard in the midst of a blowing murmur or at the beginning or end of such a murmur. A long murmur, except in the case of mitral or aortic stenosis, is usually indicative of early and comparatively slight disease and of efficient action of the heart. A short murmur may be innocent of prognostic import, but it is very frequently an indication of ruined valves and of a failing heart: it may, for instance, indicate that the orifice is so patent, say, in aortic insufficiency, that the refluent blood passes through it rapidly and with little hindrance; or in mitral disease, that the systole is brief and imperfect, the heart being on the point of breaking down.

The accent at the beginning of a murmur is chiefly observed in the regurgitant diastolic murmur of aortic insufficiency where it represents the second sound, and it is important as showing that the valves still act as a check on the reflux of blood from the aorta. It has the same significance in a minor degree as persistence of the aortic second sound—a very important fact, which will be fully explained later, in the chapter on aortic regurgitation.

It will thus be seen that the murmurs, while pointing out distinctly and certainly what valve is affected, afford also some information as to the extent of change which has taken place, though only of a vague character.

Modification of Heart-Sounds.

The heart-sounds may be modified by murmurs in various ways. A mitral systolic or an aortic diastolic murmur may accompany, replace, or follow the sound with which it is associated in the cardiac cycle. Generally speaking, when the heart-sound is distinctly heard as well as the accompanying murmur, the lesion is slight: when, on the contrary, it is entirely replaced by the murmur, the lesion is probably severe.

When a mitral murmur follows the first sound at a brief but appreciable interval, constituting a "retarded systolic" murmur, it seems to show that the valves come together accurately at first, but fail to remain in apposition throughout the whole period of the ventricular contraction. It indicates, therefore, that the changes in the valves, and consequently the amount of leakage, can only be slight. A "retarded diastolic" aortic murmur is sometimes met with, but it has not the same favourable significance.

The first sound in association with the presystolic murmur of mitral stenosis may be greatly modified. It is not replaced by the murmur, but is usually altered in character, becoming short, sharp, and high-pitched.

The Pulmonic Second Sound.

This is usually accentuated to a varying degree in mitral affections, owing to increase of pressure in the pulmonary circulation. There are two factors instrumental in causing this increase of pressure—obstruction to the outflow of blood from the pulmonary veins into the left auricle, due to the mitral lesion; and increased driving power of the right ventricle, the result of compensatory hypertrophy. The degree of accentuation of the pulmonic second sound depends on the degree of pressure in the pulmonary circulation: it will thus afford important evidence as to the amount of reflux or obstruction caused by the mitral lesion, and also as to the state of efficiency of the right ventricle. For instance, if in a case of mitral disease the pulmonic second sound is greatly accentuated, it will tend to show that the mitral lesion is one of some severity, and, as confirmatory evidence, we should expect to find hypertrophy of the right ventricle. Bronchitis, or any intercurrent lung trouble, increasing the obstruction to the flow of blood through the lungs, will tend to cause still greater accentuation of the pulmonic second sound, provided that the right ventricle does not break down under the strain. If in a severe case of mitral disease the pulmonic second sound, from being greatly accentuated, becomes feeble or much diminished in intensity, it will indicate, not, of course, that the valvular lesion is less, but that the right ventricle is beginning to fail.

The **aortic second sound** will be accentuated when the tension in the systemic circulation is high from any cause; it will be altered in intensity and pitch in aneurysm, or in dilatation of the aorta.

Reduplication of Heart-Sounds.

Reduplication of the second sound heard at the base of the heart indicates that the pulmonic and aortic valves do not close synchronously. It is of common occurrence in mitral disease, especially in mitral stenosis, when it indicates that the pressure in the pulmonary circulation has become so considerable as to cause the pulmonic valves to close before the aortic.

Reduplication of the first sound heard at the apex indicates that the two ventricles do not accomplish their systole simultaneously, owing to the fact that one is beginning to give way under extra strain imposed on it. It is not infrequently met with in advanced aortic stenosis, or in cases of high arterial tension, such as results from kidney disease; it will then show that the left ventricle is beginning to fail.

Alterations in the Cardiac Rhythm.

Intermission is less frequent than irregularity as a result of valvular disease. The particular affection in which this deviation from the normal rhythm is most liable to occur is aortic incompetence, when it indicates a faltering of the heart's action. Irregularity is very common as a result of mitral incompetence.

THE PULSE.

INFORMATION TO BE GATHERED FROM THE PULSE AS TO THE NATURE AND EXTENT OF VALVULAR LESION.

In **aortic stenosis** the artery will be somewhat small, and full between the beats; the initial percussion-wave will be slight and gradual, and the pulse-wave prolonged. This modification of the pulse is due to the fact that the blood has to pass through a narrowed orifice on its way from the left ventricle into the aorta. Hence the impact of the systole upon the column of blood in the aorta will be diminished, and more time will be required for the passage of the contents of the ventricle into the arterial system.

In **aortic incompetence** we have the well-known collapsing or water-hammer pulse; the artery is large, and empty between the beats; the pulse-wave is sudden, forcible, short, and ill-sustained, and its cessation is very abrupt.

In **mitral stenosis** the artery is small, full between the beats, with higher tension than would be expected, and the pulse-wave is long.

In all these three forms of valve-lesion the pulse is regular till the heart begins to break down.

In **mitral regurgitation**, on the other hand, the pulse is usually irregular, both in force and frequency, if the lesion is at all severe. The pulse-wave is also short, and passes the finger rapidly.

The characteristics of the different types of pulse, as brought out by the sphygmograph, are shown in the accompanying tracings.

In Fig. 4, the pulse of aortic stenosis, it will be seen that the wave is of little altitude, and has a sloping upstroke, with a rounded top and a gradual descent; that is to say, the wave is small, and attains its maximum gradually, is persistent or long, and subsides slowly. There is no dicrotic wave, as the conditions necessary for its production, great fluctuation of the blood-pressure and rapid contraction of the ventricle, are absent.

In Fig. 5, the pulse of aortic incompetence, the upstroke

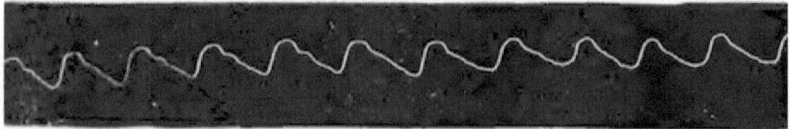

FIG. 4.—AORTIC STENOSIS.

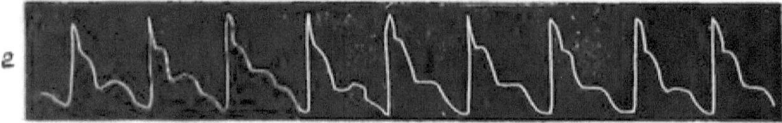

FIG. 5.—AORTIC INCOMPETENCE.

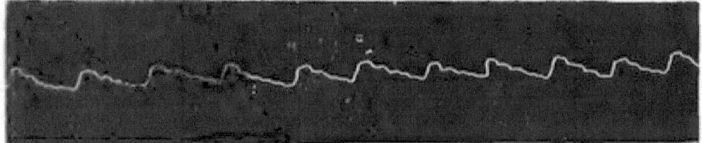

FIG. 6.—MITRAL STENOSIS.

FIG 7.—MITRAL INCOMPETENCE.

is high, perpendicular, has a sharp top, and falls rapidly; that is, the wave is large, owing to the size of the artery, is sudden and rapidly attains its maximum, is very short and rapidly falls. Dicrotism is not altogether absent, but, owing to want of the fulcrum formed by the aortic valves, it is much less marked than might be expected from the violence of the fluctuations and the rapidity of the systole.

In Fig. 6, the pulse of mitral stenosis, the upstroke or percussion-wave is short, and soon attains its maximum; the wave is long, and is slowly extinguished. Dicrotism is absent.

In Fig. 7, the pulse of mitral incompetence, the tracing shows that the pulse is very irregular both in force and frequency; the wave is short and small, and ill-sustained.

The character of the pulse, therefore, affords important information as to the nature of the lesion; while the degree in which the special peculiarities are developed, in each instance gives some clue to the extent of the valvular mischief. The pulse, however, may be modified in various ways; for instance, where aortic stenosis co-exists with regurgitation, the collapsing and sudden character of the regurgitant pulse will, to a great extent, be lost. Such modification, when present, is in itself a valuable help to diagnosis, as it enables us to say with certainty that there is real aortic stenosis, and not merely roughening of the aortic valves, either of which might be indicated by the presence of a systolic basic murmur.

Much information may thus be gained from the pulse; but the character of the pulse, even when taken in connection with the murmurs, is not sufficient to enable us to estimate the degree of obstruction or the amount of regurgitation in a given case. Further indications, however, are to be obtained, firstly, from the effects on the cavities and walls of the heart produced by the mechanical difficulties resulting from the valvular imperfections; secondly, from the evidences of obstructed circulation in the lungs or system.

CHAPTER III.

HYPERTROPHY AND DILATATION OF THE HEART: THEIR IMPORTANCE AS A MEANS OF ESTIMATING THE EXTENT OF A VALVULAR LESION—POSSIBLE OBJECTIONS TO THIS VIEW DISCUSSED—EXPLANATION OF THE WAY IN WHICH THE DIFFERENT VALVULAR LESIONS GIVE RISE TO HYPERTROPHY AND DILATATION: (1) AORTIC STENOSIS; (2) AORTIC INCOMPETENCE; (3) MITRAL REGURGITATION; (4) MITRAL OBSTRUCTION—COMPENSATION.

THE most important indications as to the extent of a valvular lesion are to be gathered from its effects on the walls and cavities of the heart, resulting in hypertrophy of the former and dilatation of the latter. These changes are due to the efforts of the heart to overcome the mechanical difficulties in the circulation occasioned by the regurgitation or obstruction to which the valvular lesion, has given rise.

Hypertrophy and dilatation must be looked upon as caused by the valvular lesion, and as affording a measure of its extent. The degree of these structural changes is ascertained by the increased area of deep cardiac dulness, by the displacement and modification of the apex beat, by the situation and character of the impulse, and by associated changes in the character and rhythm of the heart-sounds; the more pronounced these changes, the greater is the mechanical difficulty in the propulsion of the blood, and the more grave is the prognosis.

Not that a given degree of hypertrophy or dilatation, or of the two combined, is indicative in all cases of the same extent of valvular change. Each kind of valvular disease has its own special form and degree of structural change, and comparisons, to be valid, must be made between like cases; a degree of hypertrophy and dilatation, which would have little significance in aortic insufficiency, might indicate a serious amount of mitral regurgitation; and again, mitral stenosis, which has reached a dangerous degree, may be attended with less conspicuous structural change than a degree of mitral insufficiency, which gives rise neither to danger nor to inconvenience. It will be well, indeed, to exclude mitral stenosis while the general question under consideration is argued out. Numerous other modifying influences are also in operation, so that, while it is generally true that the greater the dilatation and hypertrophy, the greater the functional imperfection of the valves, the converse statement, that the less the hypertrophy or dilatation, the smaller the valvular damage, must not be taken without important qualifications. This statement is indeed usually true in the absence of symptoms indicative of interference with the circulation, but sometimes the very failure of the heart to undergo the changes required for the compensation of valvular inefficiency is what allows of the development of such symptoms, and gives rise to an early fatal termination.

It is stated by Walshe, and is undoubtedly true, that no direct ratio constantly holds good between the amount of hypertrophy and valvular change, as ascertained by post-mortem examinations. This, however, is capable of explanation, if we take into account the various circumstances that may modify the development of the hypertrophy.

Firstly. The age of the patient at the time when the lesions of the valves take place is an important factor. The enormous hypertrophy and dilatation sometimes met with are produced almost exclusively in early life, when the heart is still developing and its muscular substance is capable of active growth; later, the heart loses its adaptive capability in great measure, and thus a valvular lesion which, at the age of fifteen or twenty, would be survived with enormous hypertrophy, would at forty or fifty prove fatal with very little change.

Secondly. Time is an important element in the development of hypertrophy, which may take years to reach its maximum growth; and it is often very difficult to assign a date to the origin of a given morbid condition. For instance, of two cases in which the extent of the valvular mischief, as seen by post-mortem, is apparently the same, in one the heart may be found to be considerably hypertrophied, because the patient did not at once succumb to the injury; in the other there is perhaps comparatively little hypertrophy, because the difficulty of carrying on the circulation increased more rapidly than the power of the heart to cope with it, so that death occurred quickly.

Thirdly. We must take into account the fact that different affections of the valves have inherently and mechanically different degrees of tendency to the production of structural alterations. For instance, aortic incompetence gives rise to enormous hypertrophy of the left ventricle, whereas mitral incompetence gives rise to very moderate hypertrophy, and uncomplicated mitral stenosis to atrophy rather than hypertrophy.

Fourthly. The mode of life, whether active and accompanied by considerable muscular exertion, or sedentary and unattended by anxiety and excitement, will be an important consideration. The amount of work the heart is called upon to do will vary in each instance, and the hypertrophy will vary in direct proportion.

Again, the presence of high tension in the vascular

system will tend to increase the degree of hypertrophy to a considerable extent.

Fifthly. The period after the occurrence of the valvular change at which active exertion is undertaken, allowing or not allowing the heart to adapt itself gradually to this change, will have great influence on the condition of its walls and cavities. For example, after endocarditis, which has given rise to damage of the aortic or mitral valve, one patient has prolonged rest, care, comforts, and change of air, so that dilating influences are postponed by the rest while the heart is weak and liable to yield, and the muscular walls are enabled, by a good state of nutrition, to resist dilatation; another must return to work, or is allowed to exert himself before the heart has recovered from the effects of the illness, and its badly nourished fibres give way and permit of dilatation, which will be followed later, if the patient lives, by hypertrophy.

We have in the above conditions the quota of causation, beyond the mechanical difficulty at the valves, which explains the variation observed; and it will be evident, from these considerations, that an unvarying direct ratio between the valvular and structural changes is not to be looked for, and that its absence furnishes no valid objection to their standing in the relation of cause and effect.

MECHANISM OF CAUSATION OF HYPERTROPHY IN THE DIFFERENT VALVULAR DISEASES: ITS BENEFICIAL EFFECTS.

Aortic Stenosis.—Taking first the most simple case. If the constriction of the aortic orifice is sufficient in amount to give rise to mechanical obstruction to the flow of blood through it, there must either be increase in the propulsive power of the heart, or a slowing of the circulation; for it is obvious that the same force will not propel the same amount of blood in the same time through a narrowed orifice as through one of normal size. We find, however, that the rate of circulation is maintained, the increased power necessary to drive the blood through the narrowed orifice at a more rapid rate being gained by hypertrophy of the heart, which takes place to a degree necessary to overcome the obstruction. This is a simple illustration of the physiological law, that increased functional activity gives rise to increase of structure. If the hypertrophy did not take place, there would be a slowing and finally a standstill of the circulation.

Aortic Insufficiency.—In this lesion a certain proportion of the blood driven into the aorta at each systole returns into the ventricle through the leaking valves; consequently, in order that the same rate of circulation may be maintained, there must be either an increase in the number of heart-beats per minute, or an increase in the quantity of blood expelled by the ventricle at each contraction. Mere augmentation of the force of the systole would not answer the requirement: for instance, if five ounces of blood are driven into the aorta at each systole, and one ounce returns, the normal supply of blood would obviously not be maintained by propelling the five ounces more forcibly; what is needed is that six ounces of blood should be driven into the aorta, so that when one ounce has regurgitated into the ventricle, five ounces would still remain in the arterial system. This requirement is met by an increased capacity of the left ventricle, which is the primary compensatory change taking place in aortic insufficiency. Increased capacity is, in other words, dilatation. In this instance, therefore, dilatation of the left ventricle, which under other conditions is injurious, is actually a beneficial and conservative change when reinforced by hypertrophy, and this we shall see is a natural sequence.

It is true there is no direct provocation to hypertrophy in the shape of increased resistance to the blood-flow; but there is a call for increased exercise of force in the additional quantity of blood to be projected from the ventricle into the aorta at each systole. Further, the total internal area of the walls of the ventricle is greatly increased owing to the increase in its capacity, while the same pressure is exerted on each square inch during systole, hence the amount of force to be exercised by the muscular walls of the heart must be proportionately augmented. These two causes give rise to the hypertrophy present in so remarkable a degree in aortic regurgitation.

It must not be supposed that, because the dilatation which takes place in aortic regurgitation is compensatory and necessary, this is a sufficient explanation of its occurrence. It is produced in the following way: During diastole, the most defenceless period of the heart in the cardiac cycle, when the muscular fibres are in a state of relaxation, the ventricle is exposed to a double distending force. Firstly, the entry of blood from the left auricle and pulmonary veins; secondly, the backward rush of the blood from the aorta: the greater the amount of the regurgitation, the greater will be the distending force, and consequently the greater the dilatation of the left ventricle. This dilatation, however, is a totally different thing from the dilatation which is sometimes met with as the result of structural change and degeneracy of the muscular walls of the heart: in the latter case, the walls yield because they are inherently weak, and the effect on the circulation is disastrous, a gradually increasing stagnation from deficient *vis a tergo;* in the former case the walls yield, not because they are weak, but because they are subjected to an abnormal distending force while in a condition of relaxation and least able to resist it; ultimately, when this dilatation is backed by hypertrophy, the effect on the circulation is beneficial, since it neutralizes, more or less completely, the tendency to stagnation produced by the regurgitation.

When, however, aortic regurgitant disease is set up late in life from degenerative changes in the valves, there is frequently also degeneration and consequent weakening of the muscular substance of the heart; then, owing to the inherent weakness of the walls of the heart, the dilatation due to the aortic regurgitation will be excessive, unless the case is cut short by sudden death, and, as there is little chance of sufficient compensatory hypertrophy taking place, the condition will be one of extreme distress and danger, if the regurgitation is extensive.

Mitral Regurgitation and Mitral Obstruction.

In affections of the mitral valve the effects of the derangement of the circulation due to the lesion no longer fall mainly on the left ventricle, but primarily on the left auricle and lungs, and eventually on the right ventricle. The left auricle in mitral lesions corresponds in its relation to the current of blood to the left ventricle in aortic lesions; hence, by analogy, in mitral regurgitation, we should expect to find hypertrophy and dilatation, and in mitral obstruction marked hypertrophy of the left auricle. This is what occurs to a certain extent, but as the auricle is thin walled and poor in muscular substance, it is impossible for it to take on hypertrophy and compensate for the mitral lesions in the same way that the ventricle does for aortic lesions; it may, indeed, be distended so as to form a mere membranous sac, sometimes of enormous size. The result is that the burden of the work of compensation is thrown on the right ventricle by back pressure through the pulmonary circulation, which thus becomes the channel through which the effects of the mitral regurgitation or obstruction are transmitted from the left auricle to the right ventricle.

We find, therefore, that the right ventricle undergoes hypertrophy and dilatation as a result of mitral lesions, while the pressure in the pulmonary circulation, the connecting channel, is enormously increased.

It thus comes to pass that the right ventricle, by the additional force it gains from hypertrophy and the additional capacity it gains from dilatation, aids in supplying the left ventricle with blood, and in neutralizing the disturbing effects of mitral lesions on the circulation, since the increased pressure in the pulmonary system and left auricle will send the blood more rapidly through a constricted orifice in case of mitral stenosis, and will resist the regurgitation in mitral incompetence.

DILATATION OF THE LEFT VENTRICLE IN MITRAL INCOMPETENCE.

A certain limited amount of dilatation and hypertrophy of the left ventricle, as well as of the right, takes place in mitral incompetence, the explanation of which is as follows:—

During diastole, in consequence of the increased pressure in the pulmonary circulation and left auricle, the blood will rush through the mitral orifice with greater force and rapidity than normal; hence the pressure on the walls of the left ventricle will be greater than normal, and the muscular fibres, being relaxed and defenceless, readily yield to the extra distending force: in this way some degree of dilatation results. Then as the dilatation increases the area of ventricular wall exposed to pressure, it increases also the amount of work to be done in systole, and thus creates a demand for hypertrophy, which also takes place to a moderate extent.

In **mitral stenosis**, though there is increased pressure in the pulmonary circulation which would tend to cause dilatation of the left ventricle during diastole in the same way as in mitral incompetence, the mitral orifice, being narrowed, does not allow a volume of blood, sufficient to exert a distending force, to pass through; indeed, in severe cases, the constriction may be so great that the ventricle has never time to fill during diastole, so that it tends rather to decrease in size than dilate, and there will be no demand for hypertrophy.

In **incompetence of the mitral valve**, therefore, the amount of hypertrophy and dilatation of the right ventricle and the degree of pressure in the pulmonary circulation are among the most important indications of the extent of the lesion.

In mitral stenosis this holds good up to a certain point, but the changes in the right ventricle are not so safe a guide, since the dilatation and hypertrophy often appear to be restricted by the absence of accompanying changes of a similar character in the left ventricle.

Compensation.

It has thus been seen that dilatation and hypertrophy of the left or right ventricle or of both are a necessary consequence of valvular disease of any severity, if the patient lives, and the mechanism of their production has also been discussed. These changes in the cardiac walls are spoken of as compensatory, that is, they are changes which must take place to enable the heart to cope with the extra work thrown upon it as a result of the valvular lesions; in the case of aortic disease it is the left ventricle, in the case of mitral lesions the right ventricle, more especially which undergoes compensatory changes. Compensation is said to be established when the hypertrophy and dilatation, the former especially, have so far developed that they neutralize the disturbing effects on the circulation which the valvular lesions would otherwise produce, and enable the patient to live his ordinary life without discomfort, and without any marked symptoms.

When compensation is imperfect, the symptoms incident to the valvular disease from which the patient is suffering will present themselves more or less readily under moderate exercise or exertion, their severity varying inversely with the degree of compensation established.

For instance, a boy who is allowed to go about immediately after he has contracted a valvular lesion of some severity, and is suffering, say, from aortic incompetence, will be extremely short of breath, and incapable of walking any distance, will have attacks of severe pain in the præcordium, and perhaps fainting fits, one of which may prove fatal; whereas the same patient, if he is kept at rest till the compensatory changes have had time to develop, will be able to take moderate exercise comfortably and go about his work free from pain or respiratory distress, though he may be incapable of any prolonged or violent exertion.

In mitral disease, one of the most important guides as to the state of compensation is the size of the liver. When this organ becomes engorged and enlarged, owing to obstruction to the flow of blood from the inferior vena cava to the right auricle, it will indicate that the right ventricle is unable to cope efficiently with the increased pressure in the pulmonary circulation caused by the mitral lesion, and consequently that compensation is imperfect. A further indication will be pulsation and fulness of the veins of the neck, which may sometimes be seen to fill from below. The liver may attain an enormous size, extending well below the umbilicus, and may eventually pulsate when tricuspid regurgitation has become established.

CHAPTER IV.

SYMPTOMS IN AORTIC AND IN MITRAL DISEASE—DROPSY: DISCUSSION AS TO ITS CAUSATION.

Symptoms in Valvular Disease.—The character of the symptoms in the earlier stages differs widely in aortic and mitral disease, so that they may be divided into two groups.

1. **Symptoms in Aortic Disease.**—In aortic disease the patient is generally pale. The early symptoms are weakness, incapacity for sustained work, attacks of faintness, shortness of breath on exertion, with or without pain or a sense of oppression in the cardiac region; later, in addition to aggravation of these symptoms, amounting in respect of pain, to angina pectoris or anginoid attacks, there may be sleeplessness or terrible dreams: sometimes severe paroxysms of gasping dyspnœa occur, more rarely there may be breathlessness on lying down, so that the recumbent posture becomes intolerable.

Dropsy, as a rule, does not supervene till the mitral valve has given way as a result of the strain upon it caused by the increased pressure in the left ventricle, or has ceased to be competent from dilatation of the orifice. When once the mitral valve has become incompetent, the burden of compensation, as in mitral disease, falls on the right ventricle, and hence we may get a train of what may be called mitral symptoms in addition to the already existing aortic symptoms. Sometimes, however, dilatation of the left ventricle, without mitral reflux, will give rise to sufficient back pressure to overpower the right ventricle. The onset of dropsy may also be precipitated by kidney disease or anæmia.

In aortic regurgitation terminating by sudden syncope there has frequently been no dropsy from first to last, and often no difficulty in lying down, except during paroxysms of dyspnœa. Where there is not this sudden termination the patient may die worn out by sleeplessness and suffering, and exhausted by want of nourishment, food being taken in small quantities, or being imperfectly digested and assimilated even if eaten in sufficient quantity.

Vomiting is not uncommon, and is always a serious symptom. Effusion into the pleural cavity may take place, or an acute intercurrent pulmonary attack may prove fatal. It is comparatively seldom that death is preceded by long suffering from congestion of the lungs and dropsy.

2. **Symptoms in Mitral Disease.**—In mitral disease the early symptoms are such as are indicative of obstruction in the pulmonary circulation, and dropsy, sooner or later, is the rule. There may be pallor of the countenance; but frequently, when the valvular affection is considerable, there is more or less lividity of the cheeks, nose, and ears, seen also in the finger-nails, with capillary injection; the lips also are blue or crimson; the eyes are glistening. Sometimes, and especially in stenosis, there may be a bright, healthy-looking colour. The extremities are cold, the breath is short, the patient cannot walk uphill or against the wind without distress, especially on first starting, and the dyspnœa is conspicuous to onlookers. Very frequently, however, after walking gently for ten or twenty minutes, he regains his breath, and can ascend without difficulty a slope which at first brought him to a stand-still. Cough is common, and congestion or œdema of the lungs is easily induced. There may be attacks of

hæmoptysis due to rupture of pulmonary capillaries; these are more especially liable to occur in mitral stenosis. Later there is habitual dyspnœa, and as the effects of the disease develop themselves, dropsy comes on, beginning in the ankles as the most dependent part. When this stage has been reached, as the dropsy advances the suffering of the patient increases in proportion. The least exertion— walking across a room, rising from a chair, turning in bed— provokes a paroxysm of dyspnœa, and some degree of difficulty and oppression in breathing is experienced even during repose. The patient cannot move about, and yet is unable to lie down. In bed he must be propped up in a sitting posture, and even then may be distressed unless he lets the legs hang over the side from time to time. Very frequently he cannot remain in bed at all, but passes night and day in his armchair for weeks together. Sleep is broken or almost lost; voluntary respiratory effort is needed to supplement the ordinary reflex movement of respiration, and if the sufferer falls asleep this is suspended, dyspnœa is induced, and he wakes up out of a frightful dream struggling for breath. As dropsy increases and invades successively the thighs, scrotum, and trunk, new sources of discomfort and suffering appear, the unwieldy limbs are moved with difficulty, and no position can be found in which they are easy. Excoriations are formed at the flexures of joints, irritable eruptions appear on the legs, or the skin may crack and serum exude. Sometimes thrombosis of the veins occur, which adds to the venous stasis and gives rise to sloughing of the skin in severe cases, or gangrene of the extremity. Fluid sooner or later forms in the abdominal cavity, and may reach an amount which aggravates the suffering by its weight and by the distension of the parietes, and also, by its interference with the action of the diaphragm, adds to the respiratory difficulty. Effusion, again, may take place into one or both of the pleural cavities, and this will greatly increase the dyspnœa, the lungs being already congested and œdematous.

Dropsy.

The exact conditions which determinate the effusion of serum into the connective tissue in heart disease form an interesting subject of study, and it is important that the causation of dropsy should be clearly understood in order that the treatment of this complication may be undertaken with intelligence and confidence. The causes of dropsical effusion, as ascertained by experiment, are obstruction to the return of venous blood, loss of tone in the vessels, induced by destruction of the nerves of the part, and a watery condition of the blood. The first is demonstrated by tying a vein, and the result will be all the more decided if the lymphatic trunks are included in the ligature; the second is shown by the fact that, the two legs of a frog being exposed to the same causes of dropsy, one, having the sciatic nerve divided, will become swollen, while the other will not; injection of water into the veins has illustrated the third.

Parallel instances are furnished by disease (1) in the œdema resulting from pressure upon a vein by a tumour, or from obliteration of veins by thrombosis; (2) in the œdema of the paralyzed limbs sometimes seen in hemiplegia, and (3) in the anasarca of anæmia and kidney disease. The question is, which of these causes is chiefly in operation in heart disease: Venous obstruction, loss of tone and other changes in the arteries and capillaries, innutrition of the tissues, or deterioration of the blood?

The condition which is most certainly and conspicuously present in valvular disease is obstruction to the return of venous blood to the heart. Further, it must be borne in mind that the obstruction and damming back takes effect on the thoracic duct and lymphatic channels and capillaries as well as on the systemic veins, since the thoracic duct pours its stream into the great veins of the neck.

But that this venous obstruction is the efficient cause of cardiac dropsy has been seriously questioned. The difficulties have perhaps been best set forth by Walshe. He advances the following series of propositions :—

"1. Mitral regurgitation or obstruction, or aortic regurgitation or obstruction, may severally exist, and for a lengthened period, without systemic dropsy supervening.

2. Mitral regurgitation and aortic regurgitation may co-exist for years, and yet no dropsy occur.

3. Both of these propositions (1 and 2) hold good, whether notable hypertrophy do, or do not, exist behind or in connection with the obstruction.

4. Simple hypertrophy of the left ventricle may reach the highest point without systemic congestive effects of any kind arising.

5. Dilated hypertrophy, even of the left ventricle may last for years without any such effect ensuing, provided the dilatation be not in notable excess.

6. The heart may be in a state of advanced fatty metamorphosis, the pulse feeble and infrequent, the encephalic and respiratory functions exhibit the singular perversions attending a high degree of that disease, the entire organism betray functional languor and inactivity, and yet even the prætibial integuments fail to pit in the least under pressure.

7. Or the heart may be soft and flaccid, and the pulse persistently frequent, feeble, and irregular in force and rhythm, and yet no systemic congestions occur.

8. The natural relationships of width of the arterial orifices, and also of the auriculo-ventricular orifices, may be materially perverted, without the least systemic dropsy arising, until the closing days of life.

9. Tricuspid regurgitation, where the right ventricle is in a state of dilated hypertrophy, as shown during life, by swollen and pulsatile jugular veins which fill from below, and as shown after death by actual examination, does not necessarily produce dropsy."

These statements are, as he says, incontrovertible; he adds, "I cannot, then, see how the conclusion is to be avoided that something beyond, and in addition to, any one or any group of the cardiac conditions referred to, is required in order as a matter of necessity to entail the occurrence of dropsy.

" And again, the existence of some active cause beyond, and independent of, the heart, is further shown by the facts, that there is no direct relationship between the amount of heart disease and of dropsy; that dropsy comes on suddenly, sometimes from extraneous causes, the state of the heart remaining, as far as ascertainable, in precisely its previous condition; and that dropsy diminishes and increases, comes and goes, either spontaneously or through the influence of treatment, while the organic changes in the heart remain permanent and unmodified."

While admitting the truth of all Walshe's propositions, it should be added that they relate to exceptions, and the great fact of the association of heart disease and dropsy remains. Were a few exceptions, however unaccountable, to be allowed to invalidate a general law or deductions from general experience in medicine, few of the clinical or therapeutical conclusions in which we are accustomed to place confidence would stand. But most of the difficulties advanced in the propositions quoted are capable of explanation. For the conclusions deduced by Walshe are only valid on the supposition, that the effect on the circulation is exactly the same when apparently similar conditions of valves, cavities, and heart-walls are present. But such is not the case; the influence on the circulation generally, and especially on the return of blood by the veins, may vary greatly when very similar affections of the walls and valves are present. Let us suppose that a patient suffering from any one of the combinations of heart disease enumerated, and who up to a given moment has no dropsy, begins to suffer from constipation or flatulent dyspepsia, or becomes a prey to some serious anxiety. Without any change in the organic condition, the functional disturbance or depressing influence may seriously derange the compensatory adjustment which has so far been sufficient, and may thus permit of the appearance of dropsy. Here again removal of the cause might be followed by recovery from the dropsy, the state of heart remaining the same. Or, the patient is exposed to cold and contracts severe bronchial catarrh, the additional resistance caused thereby in the pulmonary circulation is more than the right ventricle, already enlarged to the utmost of its power, is capable of coping with, and the back pressure in the systemic veins is augmented to the point at which serum exudes into the tissues. Let us suppose, again, that the heart is in exactly the same condition in two individuals, but one of them is exposed to hardships, while the other is surrounded by every care. One will have dropsy, the other not.

It must be remembered that we have only certain names and phrases at our command in order to describe the changes in the heart; and differences which defy description, such as variations in the relative proportions of cavities, walls, and valve-lesions, may affect the equilibrium of forces in the circulation in an important degree. We can indeed only trace coarse and conspicuous compensations, such as hypertrophy; we cannot estimate the relative contractile efficiency of the muscular fibres; other rectifying processes may also be at work which escape us. We need only allude to the hypertrophy in the muscular coats of the arteries, described and called attention to by Sir George

Johnson, which plays so important a part in Bright's disease.

Another explanation of some apparent anomalies is, that the mechanical condition necessary for the production of dropsy is such obstruction to the return of blood to the heart, as leads to increases of pressure in the venous radicles and in the capillaries. The cause of the effusion is not simply languid movement of blood in the capillaries, and a tendency to stasis, but a disturbance of the accord between the outflow from, and injection of blood into, the systemic circulation. If from any cause the pressure is not supplied by the ventricle from behind, the hydrostatical condition for exudation of serum into the tissues is wanting. Changes in the heart, then, which apparently or actually add to the gravity of the case by diminishing the force or efficiency of the heart's systole, may by lessening the *vis a tergo* tend to postpone the particular symptom of dropsy.

When there is extensive valvular disease together with considerable secondary change in the ventricles, we have a state of unstable equilibrium in the circulation. In health or disease, the heart is perpetually called upon to adjust itself to new conditions; changes of temperature, by contracting or relaxing the cutaneous arteries and capillaries, increase or diminish the resistance to the passage of blood through them, and the degree of force required to overcome this resistance; other influences affect the abdominal circulation, the amount of blood circulating in the mesenteric and other arteries, or accumulating in the large veins; the pressure of muscles in action drives the blood from the limbs to the right side of the heart more or less rapidly, and this has to be forwarded through the lungs; excitement accelerates, or depressing emotions lower the rate and vigour of the heart's action. In the normal state it responds readily to the calls made upon it; but in disease the adjustments are effected with difficulty, and under comparatively slight provocation may fail altogether, when the balance of compensation will be overturned, and symptoms warded off up to this point will begin to develop themselves.

Walshe does not, of course, deny the causation of dropsy by heart disease, and indeed says, " We must not run into the opposite and equally erroneous extreme of wholly ignoring the direct influence of organic changes of the heart and its orifices ; " but he appears to limit its operation to the establishment of a difficulty in the systemic circulation, and attributes to other causes the actual and direct production of dropsy. This influence he looks upon as compound ; *i.e.* an impoverished state of the blood and a conceivable varying density of texture of the walls of the vessels, attaching, however, most importance to the blood condition, which he believes to be induced by some other agency than the imperfect circulation resulting from the heart affection. But while the influence of anæmia is admitted to the full and has been already specially considered, the deteriorated state of the blood, together with the changes in the capillaries and tissues, can only be regarded as tendencies co-operating with those originating in the impeded circulation. An effective agency in the production of the blood and tissue changes referred to is a retarded circulation. The sluggish stream can neither furnish the digestive secretions in due quantity and of proper quality, nor take up the nutritive materials with normal rapidity, nor can the assimilating organs, impeded in their functions by chronic congestion, effect perfectly the further changes necessary to the formation of healthy blood. The tissues also, permeated only by a slow current of unhealthy blood, can neither obtain sufficient matter for their renewal, nor get rid of the products of waste, and consequently fall into a state of degeneracy. It might be argued, that since dropsy does not usually supervene until disease of the heart has lasted over a considerable period, there will have been time for deterioration of the blood and vessels, and that this is a necessary intermediate cause. Its intervention in a large proportion of cases is not questioned, but that it is a necessity cannot be admitted.

A very common occurrence in heart disease seems to me to demonstrate the mechanical causation of dropsy. A patient suffering from valvular disease, more or less perfectly compensated by hypertrophy, has œdema of the ankles after sitting up and moving about, which disappears on lying down for a sufficient time, and the fluid, so far as can be ascertained, has not merely been diffused in the cellular tissue generally, but has been absorbed. After a few days' rest in bed, the swelling will frequently not reappear for some time, showing that resorption of the fluid has taken place.

Or, on the other hand, he exerts himself imprudently, either in a single violent effort, or by moderate work and exercise too long sustained, and begins at once to suffer from breathlessness or dyspnœa, which is soon followed by dropsy. We can scarcely suppose that under these circumstances there has been such a change in the blood or tissues as to determine the effusion; clearly the determining influence is the increased derangement of the circulation; very frequently an increase of dilatation of one or more of the cavities of the heart is recognizable, and whether this is so or not, it is clear that the equilibrium between the valvular affection and the compensatory changes in the heart-walls has been overthrown. When, again, such a patient, after exposure to the cold, suffers from bronchitis or pulmonary congestion rapidly leading up to dropsy, it is surely the increased obstacle to the transit of blood through the lungs, and the consequent backward pressure in the right auricle and great veins, which give rise to the exudation of serum from the venous radicles and capillaries. There may, in the latter case, be pyrexia, it is true, which will affect the blood and dilate the peripheral arteries, but we see no such phenomenon as dropsy from this cause independently of venous obstruction.

The most convincing evidence, however, that the chief cause of dropsy is back pressure in the veins, and not deterioration of the capillary walls or any blood condition, is the rapid resorption of the fluid which may take place, and the fluctuations which are observed in œdema. The structural condition of the capillaries, or the state of the blood, cannot be quickly modified, whereas the obstruction to the venous return may vary from hour to hour.

Clinical experience is in accord with the view that dropsy in heart disease is the result of obstruction to the return of venous blood to the right side of the heart. Mitral insufficiency is the valvular affection which of all others most constantly gives rise to dropsy, and all the stages of backward pressure from the original seat of arrest in the circulation are easily followed—pulmonary congestion, hypertrophy and dilatation of the right ventricle, and tricuspid reflux. Obstruction to the pulmonary circulation by emphysema and chronic bronchitis is another common cause of dropsy, the mechanism being the same.

Mitral stenosis is not by any means so liable to be followed by general dropsy as insufficiency. This unexpected variation is not explained by any difference in the effect on the right side of the heart. Hypertrophy and dilatation and tricuspid insufficiency are equally marked, as is also back pressure in the great veins indicated by jugular pulsation and engorgement of the liver. The probable explanation is that the diminished output from the left ventricle does not allow of sufficient pressure in the capillaries to give rise to effusion of serum.

Aortic insufficiency and obstruction are not primarily

attended with dropsy, but, as has already been mentioned, they may give rise to it eventually by damming back the blood in the left ventricle, and thus producing back pressure in the same way as in mitral disease.

The effects of primary disease of the muscular walls of the heart are not so easy to trace, but the clinical facts are that, while dilatation is very frequently followed by dropsy, fatty degeneration rarely gives rise to it. In both there may be a tendency to stasis in the systemic veins and capillaries, but in degeneration, the *ris a tergo* in the arterial circulation, which is necessary for the production of pressure in the capillaries and effusion through their walls, is lacking.

CHAPTER V.

ÆTIOLOGY OF VALVULAR LESIONS—ACUTE ENDOCARDITIS—CHRONIC ENDOCARDITIS: DEGENERATIVE CHANGES—RUPTURE OF VALVE—DILATATION OF THE ORIFICE, SECONDARY TO CARDIAC DILATATION—CONGENITAL LESIONS OF VALVES.

Acute Endocarditis.—This, by far the most common cause of valvular lesions, is usually an incident of acute or subacute rheumatism. It may also develop in the course of scarlet fever, chorea, and acute nephritis, or more rarely during an attack of measles or some other of the acute diseases of early life. In childhood and early adolescence, endocarditis is almost the sole cause of valvular affections. It may also be the origin later in life; but first attacks of rheumatic fever are less common in adults, and are less liable to be complicated by endocarditis; whereas in children, though the articular evidences of rheumatism are, as a rule, slight, the heart comparatively rarely escapes damage.

Acute inflammation of the endocardium, as of other structures, is a definite process. The valves and tendinous cords are swollen and infiltrated, a certain amount of lymph is thrown out in and upon the endocardium covering them. This lymph, when the acute stage is over, undergoes organization into fibrous tissue, contracting in the process, and giving rise to thickening, puckering, and stiffness of the valves, perhaps to adhesion between them at the angles of contact, and to thickening, shortening, and sometimes to rupture of the tendinous cords. The formation of fibrous tissue comes at length to an end, and the valves are left in a damaged and imperfect functional condition. The lesion, however, once formed, with some exceptions, is not progressive in the absence of fresh attacks.

This fact, that a valvular lesion caused by acute endocarditis is for the most part stationary, makes an attack of rheumatic fever so important a date in the history of heart disease.

The principal exception to the statement that lesions left by acute endocarditis are stationary is when, by adhesion between the flaps of the mitral valve, stenosis of this orifice is produced. This lesion frequently appears to be progressive, doubtless owing partly to further contraction of the cicatricial fibrous tissue formed round the mitral orifice as a result of the inflammatory process, partly to extension of the adhesion between the valvular curtains.

Chronic Endocarditis and Degenerative Changes.

Chronic endocarditis is mostly met with during or after middle life, when it can scarcely be separated clinically from degenerative change, known as atheroma. It is rare in early life, but there may be cases in which valvular disease gradually creeps on in patients suffering from subacute or chronic rheumatism, or in young subjects of inherited gout.

Among the best-known causes of chronic valvular changes are gout and kidney disease, and it was long accepted as the explanation of their influence that the gouty or uræmic poison, acting as an irritant upon the endocardium, set up inflammation. Valvular changes of like character are extremely common when neither gout nor renal disease exists, and a mode of causation has been pointed out, which not only applies to such cases, but explains also the influence of certain occupations—that of the miner or collier, the hammerman, the soldier, etc.—in their production. It may almost, indeed, be said that the causation was first clearly understood by means of these occupations, through the investigation and reasoning of Clifford Allbutt, the late Milner Fothergill, and others. The cause is undue pressure within or upon the aorta. In the occupations mentioned, the pressure is created by effort in a constrained position, or by work of such a character as to require prolonged closure of the glottis and fixation of the chest by holding the breath. The compression exercised upon the contents of the chest necessarily takes effect upon the column of blood in the aorta, and gives rise to a more forcible and violent closure of the valves at its root. At the same time, when there is high pressure in the aorta, greater force will be required to drive out the blood from the ventricle, and this puts undue stress upon the mitral valves and their tendinous cords. A violent strain seventy or eighty times a minute, in the long run, sets up a chronic irritative process, and at the same time probably interferes with the circulation in the valves, thus favouring degenerative changes. Similar effects are produced by the high arterial tension habitually present in gout, kidney disease, and other states of system. The strain on the valves may not be as violent as in severe exertion of the character alluded to, but it is more continuous, goes on in the intervals of work as well as during effort, and by night as well as by day. From the extreme frequency with which high arterial tension is met with at and after middle life, it is far more important as a cause of valvular change than occupation.

Rupture of a Valve.

Rupture of a valve is of rare occurrence. It is usually the aortic valve which suffers, and, except as a result of direct violence, the rupture will nearly always have been preceded by disease in its structure, and by the cause of such disease, high arterial tension. The rupture usually takes place during some severe muscular exertion, and is attended by great pain in the region of the heart. It may be a complete or partial rupture of one of the cusps of the aortic valve, or may be a mere slit in it.

Mitral Valve.—One or more tendinous cords of the mitral valve may be found ruptured when this valve has been the seat of disease; but this occurrence can rarely be dated by any recognized accession of symptoms, though it must give rise to a considerable increase in the derangement of the circulation.

Dilatation of the Orifice of a Valve.

The Aortic Orifice.—Dilatation of the orifice may upset the functional efficiency of a valve while the valve itself has not undergone any material change. The aortic orifice is less liable to be stretched and enlarged than the mitral, as would be expected from the difference in the structures surrounding the two. It is only comparatively late in life that the strong fibrous ring at the root of the aorta ever yields, and it will then only take place as part of a general dilatation of the first part of this vessel. The usual result is more or less insufficiency or regurgitation; but, paradoxical as it may appear, there may actually be obstruction as well, the valves being stretched across the dilated mouth of the aorta so tightly that they cannot fall back.

The Mitral Orifice.—Mitral insufficiency from failure of fairly healthy valves to close the orifice is not uncommon.

Sometimes the orifice is very much enlarged, admitting four or five fingers instead of three. Frequently, however, there is no stretching of the opening, and the valves themselves are unaffected, although during life there has certainly been regurgitation. Various explanations of this fact have been offered. It has been supposed that irregular action of the papillary muscles interferes with the accurate adjustment of the valvular curtains; and, again, that the papillary muscles, being carried by dilatation of the ventricle so far from the valvular ring that the tendinous cords are dragged down too far, do not allow the margins of the valves to meet during the systole.

The real cause of the imperfect closure of the valves is that originally suggested by Donald McAlister.[*] He pointed out that an important factor in the valvular mechanism is the active contraction of the mitral orifice with the systole of the ventricle. We are accustomed to imagine that this opening, like the aortic, is surrounded by a strong fibrous ring, maintaining its form and patency under all circumstances; but this is not the case, as was shown by Sibson in dissections and specimens, which I had the pleasure of assisting him to prepare; for it was demonstrated that the mitral ring of fibrous tissue, taking its origin from the central fibro-cartilage of the heart, gradually thins out till at the opposite side of the orifice it is practically non-existent. There is nothing, therefore, to keep the mitral orifice rigid, or prevent it from altering its shape; hence we find that, during systole, the transversely circular fibres of the cardiac muscle, by their contraction, cause a narrowing of the mitral orifice, and partially close it independently of the valves. Such a closure is in effect a part of the general obliteration of the ventricular cavity during systole. When, therefore, there is dilatation of the ventricle, this active constriction of the auriculo-ventricular opening is

[*] *British Medical Journal*, August, 1882.

imperfectly performed, and the valve fails to cover the whole area of the orifice.

Mitral regurgitation caused in this way, may occur at any period of life, and under very different conditions. The cardiac dilatation may be the result of imperfect nutrition in advancing years, or of debilitating influences, or it may be induced by acute disease, such as enteric fever or acute rheumatism. Perhaps the most common and important cause is anæmia; it is most important because of its frequency in young adults, and because, though curable, it is yet liable to be rendered permanent by imprudence and neglect. Dr. George Balfour has rendered good service by calling attention to this condition under the head of "Curable Mitral Regurgitation." Now, although weakness of the cardiac muscular fibres, or want of energy in their contraction, is perhaps an essential condition in the production of dilatation, another factor plays an important part, namely, high arterial tension. We should not expect anæmia to be attended with any such state of the circulation, but this is very frequently the case in the ordinary type.

Pernicious anæmia, however, is less commonly attended with high arterial tension, perhaps because of the frequent intercurrent attacks of pyrexia; for pyrexia, as is well known, relaxes arteries and lowers the arterial tension.

It may be remarked in passing, that the low arterial tension which accompanies pyrexia probably renders the occurrence of dilatation of the ventricle much less frequent in acute disease than we should expect as a result of the debilitating effect of fever on the muscular walls of the heart.

Mitral incompetence may, further, be a secondary consequence of aortic incompetence, through the dilatation of the left ventricle. But, unless cardiac debility have a share in the production of this dilatation, a comparison between it and the dilatation just spoken of is illusory. The one is

F

the result of asthenia, the other of over-distension of the ventricular cavity; in the one, the contraction of the ventricle is never completed; in the other, it is carried through energetically.

Congenital Valvular Affections.

Lesions of one or more of the valves of the heart are sometimes found at birth. They may be the result of a congenital malformation, or of endocarditis occurring while the fœtus was in utero. In the former case the pulmonic valve is most commonly at fault, the lesion being usually constriction of the orifice, and frequently the valvular defect is associated with a malformation of some other portion of the heart. In the latter case the valves of the right side of the heart usually suffer, but those of the left side may also be attacked, the aortic more commonly than the mitral.

CHAPTER VI.

PROGNOSIS IN VALVULAR DISEASE—THE NATURE OF THE LESION : THE RELATIVE DANGER ATTACHING TO EACH PARTICULAR LESION—SUDDEN DEATH : THE VALVULAR DISEASES IN WHICH IT IS LIABLE TO OCCUR—THE EXTENT OF THE LESION—THE STATIONARY OR PROGRESSIVE CHARACTER OF THE LESION AS INFLUENCING PROGNOSIS.

Prognosis.—In cases of heart disease, prognosis is of special importance, since a patient who knows that he is suffering from some affection of the heart immediately dreads the worst. It is always necessary to allay his fears as far as possible, as these in themselves tend to aggravate the danger attending the disease. In some cases this may be done with absolute confidence; in others, these apprehensions may be only too well founded ; in others, again, the issue may be uncertain and may be dependent on other conditions than the state of the heart. It will be most important in all instances that the medical attendant should form a definite idea of the probable effect the heart disease will have in shortening life, so that he may guide the patient and his friends in making arrangements on which the welfare of the family may depend.

The following are the points which should specially be considered in regard to prognosis :—

1. The valve affected and the relative danger attaching to the particular lesion.

2. The extent of the lesion.
3. The stationary or progressive character of the lesion.
4. The degree of soundness and vigour, functional and nutritional, of the muscular substance of the heart and of the tissues generally.
5. The age of the patient.
6. The family history, especially in regard to whether there is any hereditary tendency to heart disease.
7. The habits and mode of life of the patient.
8. The presence or absence of other diseases—such as anæmia, bronchitis, renal affections—as complications.

1. THE RELATIVE DANGER ATTACHING TO THE DIFFERENT VALVULAR LESIONS.

According to Walshe, the valvular affections stand in the order of relative gravity as follows: tricuspid regurgitation, mitral regurgitation, mitral constriction, aortic regurgitation, pulmonary constriction, aortic constriction. Tricuspid regurgitation, however, as has already been said, is rarely primary, but usually occurs as a result of obstruction to the transit of blood through the lungs either from disease in these organs, or from valvular disease in the left side of the heart. It is therefore an effect of serious valvular lesions of the left side of the heart and can scarcely be regarded as a cause of the fatal termination.

My own experience would lead me to modify Walshe's arrangement somewhat, and to give this order of relative danger: aortic incompetence, mitral stenosis, aortic stenosis, mitral incompetence. Aortic incompetence is most rapidly fatal when it comes on late in life at a period when compensatory hypertrophy is established with difficulty, more especially when due to degenerative change in the valves. In childhood and early adolescence, mitral stenosis is often more serious than aortic incompetence, owing to the progressive nature of the lesion and to imperfect development of the left ventricle. It is, however, difficult to estimate with any accuracy the relative danger of different valvular lesions, as so many other factors must be taken into consideration in prognosis.

Sudden Death.

One of the first questions to be discussed is the liability to sudden death in heart disease. In the mind of the general public, disease of the heart and sudden death are so closely associated that the mention of the one immediately suggests the other, and in a nervous patient, a pain or a sense of weight or oppression in the cardiac region, easily exaggerated by the concentration of his attention upon the heart, will make him think the end is near.

It is therefore of the greatest importance that we should know with certainty, in what form of heart disease sudden death is liable to occur, and be able in cases where no such danger exists to say so with confidence. It must be understood that the sudden death under consideration is such as is meant by the familiar phrase "dropping down dead," with little or no warning, the individual having been up to the moment in apparent health, or so far well as to be able to go about his duties, or at any rate not suffering from dropsy or other serious symptoms of cardiac embarrassment.

In all forms of heart disease, when the effects on the circulation have become very decided, and such symptoms as engorgement of the lungs, with dyspnœa, dropsy, effusion into the pleural cavities, albuminuria, have set in, the final struggle may come on abruptly and end speedily. This, however, is not the mode of death which is the special dread attending heart disease.

Walshe, in speaking of the different forms of valvular lesions, says, that only one causes sudden death, namely, aortic incompetence.

In a paper read before the Harveian Society in 1866, I gave as the conclusion at which I had arrived that "sudden death is a contingency which may almost be left out of consideration in valvular disease, except in aortic regurgitation." This would still express very nearly my individual experience. As will be shown in the part of this book on the structural changes in the wall of the heart, there are other conditions more likely to give rise to sudden death than aortic regurgitation. Here, however, we are considering only the valvular lesions.

As individual experience is not altogether a trustworthy guide, some years ago Dr. Sidney Phillips, at my request, went through the post-mortem records of St. Mary's Hospital of four hundred cases in which the heart had undergone marked change, in order to see what light they threw on the question of heart disease and sudden death. Of these 151 were cases of valvular disease, 204 examples of changes in the wall of the heart. Of aortic regurgitation there were thirty-eight cases. Three were brought to the hospital dead, a fourth died in the hospital suddenly during convalescence from acute rheumatism. In six more the final symptoms came on abruptly and were rapidly fatal, one dying within twenty-four hours of his admission to the hospital, another within two days. There were eleven examples of aortic stenosis without one sudden death in the sense of the patient being overtaken by death while in apparent health or free from symptoms arising from heart disease.

Of mitral stenosis there were fifty-three instances. One patient only was brought in dead, a young girl who was picked up in the street. From the condition of the lungs there was no doubt that she must have been suffering severely from symptoms due to the heart affection.

Of mitral insufficiency there were forty-nine cases. Of these two may be said to have died suddenly, but both had serious symptoms and were under treatment in hospital, and in both the pericardium also was universally adherent.

In three more a final attack of dyspnœa set in abruptly and proved rapidly fatal.

These numbers are not given as representing with anything like accuracy the proportion of sudden deaths in the different forms of valvular disease, but they afford confirmation of the opinion formed from personal experience that aortic insufficiency is the only form of valvular disease attended with danger of sudden death.

The Extent of the Lesion.

In a previous chapter the indications by means of which the extent of the lesion may be estimated, have been discussed. A certain limited amount of information on this head is obtained from the character of the murmur. Further help is derived from the pulse, and still more from the amount of hypertrophy and dilatation which the heart has undergone in consequence of the lesion. A valvular murmur, accompanied by dilatation or hypertrophy or both, is attended with greater danger than a similar murmur not so accompanied; not, however, because the hypertrophy and dilatation add new elements of danger, but because the valvular change causing the murmur has given rise also to mechanical difficulty when these changes are present, whereas their absence shows that it has given rise to no serious obstacle to the circulation.

It must be understood, however, that when hypertrophy and dilatation are taken as a measure of the valvular lesion it is in the absence of any evidence of functional inefficiency. Where, in addition to the disease of the valves, there is degeneration of the muscular substance of the heart, this will give rise to further dilatation. Still, however, the amount of dilatation may be taken as expressing the relation between the mechanical difficulty and the power of the heart to cope with it. Naturally the more extensive the lesion, as estimated by the changes in the heart, the more serious the prognosis, because the compensatory balance has been established with difficulty and is more easily upset.

The Stationary or Progressive Character of the Lesion as influencing Prognosis.

A given state of valve existing—a certain degree of obstruction or incompetence—it will be obvious that the future of the patient will be very greatly influenced by the question whether the morbid process which has damaged the valve, is still in progress or has come to a standstill. In the one case, where the change has reached its maximum, we know what we have to deal with—the hypertrophy and dilatation give an approximate idea of the functional imperfection; they compensate it or they do not, and the prognosis varies accordingly: in the other there can be but one course and issue, a gradual or swift aggravation of symptoms, unless indeed sudden death interferes.

Such a difference exists arising out of the character of the pathological process by which the valve is affected. Speaking generally, the question is whether the valvular change has had its origin in an acute inflammatory attack or is the result of a chronic degenerative process. Fortunately the distinction is for the most part easy.

In the chapter on the ætiology of valvular disease the various causes of lesions of the valves and orifices have been enumerated and briefly discussed. They are, acute endocarditis, chronic endocarditis, and degenerative changes in the valves, rupture of valve, dilatation of the orifice.

Acute Endocarditis.—When the valvular lesion can be traced definitely to an attack of acute endocarditis, there is this favourable element in the prognosis, that the lesion once established is not progressive. If the lesion was slight it remains slight, and does not increase in severity after the endocarditis has subsided; consequently the dilatation and hypertrophy of the heart will be small, if indeed they are appreciable; if the lesion is extensive, the dilatation and hypertrophy which will follow are proportionate. Hence we are able to arrive at a definite conclusion as to the extent of the lesion, and give an approximate idea as to the probable effect it will have in shortening life, and say how far exercise and exertion can be allowed. The only exception to be made will be in cases of mitral stenosis, where the orifice may become gradually further narrowed after the initial damage has been done, owing to cicatricial contraction of the inflammatory products.

Chronic Endocarditis.—The fatal feature about this chronic inflammatory or degenerative change in the valves is, that once begun it is inevitably progressive. The heart and system may have accommodated themselves to a degree of obstruction or regurgitation, but this does not remain the same; slowly or rapidly the valvular lesion will increase, and with it the obstacle to the due transmission of the blood through the heart. This takes place at an age when the heart is little able to adapt itself to change or to meet the increasing difficulty, and is, moreover, itself liable to structural decay; further, the cause which has worn out the valves, the excessive resistance in the peripheral circulation giving rise to unduly high tension in the arterial system, is probably still in operation.

Before, however, we take so serious a view of any given case in which a valvular murmur has been developed late in life, and accept it as necessarily indicative of progressive disease, we ought to make sure that it is due to degenerative changes, and not to mere roughening of the valve. To prove the former we ought to have evidence of actual damage to the valve, either in the shape of symptoms traceable to functional inefficiency or of modifications of the physical signs. It is common in elderly people to find a mitral or aortic systolic murmur develop, which persists for years without any symptoms of heart disease, due to a slight roughness or rigidity in the case of the aortic valve which sets up vibrations; or, in the case of the mitral valve, to a little thickening or want of pliability which prevents an accurate apposition of the flaps of the valve, but does not allow of any appreciable regurgitation.

Rupture of Valve.

Rupture of a valve is always a very serious lesion. The valve affected is usually the aortic, and the sudden and severe strain on the heart, which has no time to accommodate itself to the altered conditions of the circulation, leads to extreme dilatation of the left ventricle and the consequent onset of severe symptoms when the rupture of a cusp is complete. The patient may be seized by a syncopal attack, which will at once prove fatal. He may, however, appear to progress favourably for some time, but will rarely recover sufficiently to be able to go about again as usual, for the accident usually takes place at a time in life when degenerative changes are already beginning to take place in the walls of the heart, and it is incapable of undergoing sufficient hypertrophy to compensate for the valvular lesion. Hence the ultimate cause of death, if the patient survives, is a gradual stasis of the circulation, the premonitory indications being a rapid increase in the size of the liver, with dilatation of the right ventricle following on that of the left, and later the onset of œdema of the extremities. It may be some weeks or months after the accident before the fatal termination ensues.

The early symptoms, however, are not necessarily so severe.

In a case that was under my care at St. Mary's Hospital in 1893, the patient, a man aged fifty-two, walked up to the casualty room, complaining that he could not use his right arm, and saying that he thought he had sprained it while lifting a heavy weight. The history of the accident was that he was carrying a heavy bag on his shoulder; and when throwing it off his shoulder he tried to prevent it falling with too much violence on to the ground by catching hold of it with his right hand. He then felt a sudden pain in his arm and chest, and was violently sick. He walked up to the hospital, apparently without much discomfort. He was somewhat pale, but was not suffering from dyspnœa, and complained chiefly of his right arm. On examination the right radial pulse was found to be absent owing to thrombosis of the brachial and axillary arteries. The pulse in the left radial was 98, regular in force and frequency, slightly collapsing; respirations, 18.

On examining the heart the apex beat was found to be in the fifth space, just inside the nipple line; a faint systolic and diastolic murmur were audible at the apex, and a diastolic murmur was also heard at the aortic cartilage and down the sternum. The liver was not enlarged.

He remained in hospital from April 22 till May 13, when he left at his own request. During this time no serious symptoms had developed, and he had been up and walking about the ward without any discomfort beyond a little shortness of breath.

Rather more than a fortnight later, on May 30, he came back to the hospital complaining of extreme shortness of breath and swelling of the legs. The pulse was then 116, the respirations 34. The legs were œdematous, and the face was very pale and anæmic. On examining the heart it was found to be much dilated, the area of cardiac dulness extending up to the third rib above and inwards to the middle of the sternum, outwards for half an inch outside the nipple line.

The apex beat was not visible or palpable, but the sounds were best heard in the seventh space, half an inch outside the nipple line. A systolic and long soft diastolic murmur were audible at that point, and the diastolic murmur was heard down the sternum. The aortic second sound was almost entirely replaced by the murmur, and the first sound was very feeble. The liver was much enlarged, extending down to within two inches of the umbilicus, and was pulsating.

Rest and treatment failed to improve his condition, which got steadily worse, the symptoms increasing in severity till he died on August 9, some ten weeks later. At the autopsy it was found that the anterior flap of the aortic valve was ruptured, and that there was extensive incompetence. There was considerable dilatation of the cardiac cavities, and little or no compensatory hypertrophy.

Dilatation of the Orifice.—The aortic orifice is less liable to dilatation than the mitral, owing to the strong fibrous ring surrounding it. When dilatation does occur, it usually takes place comparatively late in life as a part of a general dilatation of the aorta due to degenerative changes in the walls of the vessel or its orifice: it is therefore progressive in character and the prognosis is unfavourable, more especially as these same degenerative changes may lead to the production of aneurysm.

The Mitral Orifice.—Speaking generally, mitral regurgitation, established by means of dilatation of the left ventricle, in the young, or as a result of acute febrile conditions or anæmia, is stationary and, under favourable conditions, curable. When it is secondary to aortic disease, the prognosis necessarily merges in that of the primary lesion. When it occurs as a result of protracted high tension or kidney disease, it may be temporarily curable by suitable treatment, but will be liable to recur.

In later life and old age it may be gradually and imperceptibly established without any obvious cause for it, and it is difficult to diagnose between dilatation of the orifice and changes such as gradual thickening and contraction of the valves. It is in such cases, however, very slowly progressive, as a rule.

CHAPTER VII.

PROGNOSIS CONTINUED—AGE, SEX, HEREDITY—EFFECTS OF HIGH ARTERIAL TENSION—HABITS AND MODE OF LIFE OF THE PATIENT—ANÆMIA—THE CIRCUMSTANCES UNDER WHICH PROGNOSIS MAY HAVE TO BE MADE: (1) IMMEDIATELY AFTER ACUTE ENDOCARDITIS; (2) WHEN THE VALVULAR LESION IS SLIGHT AND HAS GIVEN RISE TO NO STRUCTURAL CHANGES IN THE HEART; (3) WHEN COMPENSATORY CHANGES HAVE TAKEN PLACE BUT NO SYMPTOMS OF EMBARRASSMENT OF THE CIRCULATION ARE PRESENT; (4) WHEN SYMPTOMS OF FAILURE OF COMPENSATION HAVE SET IN; (5) IN ADVANCED VALVULAR DISEASE WHEN SEVERE SYMPTOMS OF CARDIAC FAILURE HAVE SUPERVENED.

Age.—Little need be said with regard to age as affecting the prognosis of heart disease. Late in life, degenerative processes are almost invariably in operation, and the hypertrophy which is needed, in order that the effects of valvular affections may be neutralized, is established with difficulty; moreover, compensatory hypertrophy, which has served its purpose for twenty or thirty years, may at the end of this time be undermined by fatty or fibroid degeneration of the cardiac walls.

Childhood.—In early childhood, the outbreak of rheumatic endocarditis is always a matter of grave prognostic significance: firstly, because it is so frequently accompanied by pericarditis, which may prove fatal at the time of the attack or leave the heart permanently hampered and disabled by pericardial adhesions; secondly, because, even though the heart escapes serious damage from the first attack of either peri- or endo- carditis, both are extremely liable to recur. Further, it would appear that when a valvular lesion of some severity is established, the heart cannot both answer the demand for hypertrophy and keep pace with the active growth of this period of life, so that the child is liable to remain small and stunted with clubbing of the fingers and toes, and to be generally backward in development.

Sex.—It is a remarkable fact that mitral stenosis is very much more common in women than in men. The post-mortem statistics previously referred to, show that out of 53 cases of mitral stenosis, 38 were in females, and only 15 in males, which is approximately a proportion in accord with the experience of most observers. No satisfactory explanation of this predominance has yet been given.

On the other hand, aortic insufficiency is more frequently met with in men, which again is borne out by the same statistics, as out of 36 cases, 30 were males, and only 6 females. This appears to be explained by the occupation and mode of life of men; in my own experience, however, aortic regurgitation has been much more common in boys than in girls at a period of life long before the influence of occupation would begin to operate.

When valvular disease has been established in childhood, girls are, according to my experience, more likely to break down at the trying period of puberty than boys, and, speaking generally, the compensatory changes in the heart walls are less perfectly effected in the female than in the male.

Hereditary Tendencies.—In no class of cases is it more necessary to inquire into the family history than in diseases of the heart. It is more particularly in affections of the muscular walls that a family tendency to heart disease is seen, which may take the form of primary fatty degeneration, or of fibroid change secondary to high arterial tension. I have known a family in which three out of four brothers died suddenly before reaching the age of fifty-five from disease of the heart or aorta, and other examples almost equally striking; similar histories must be known to most medical men of large experience. It is not, however, only a special liability to structural degeneration of the heart at a certain age which is important. In any valvular affection, and at any stage, the constitution of the patient is an element in the prognosis which must be kept in view. We have to take into account, not simply the cardiac lesion and the functional derangement of the circulation caused thereby, but also the behaviour of the system under this disturbing influence and its power of endurance, and in a short-lived family we cannot have the same confidence in the tissues as in a family noted for longevity. More than once in my own experience, a prognosis based upon the state of the heart has been falsified through failure of general constitutional power.

Effects of High Tension in the Circulation.

High tension in the arterial system, frequently a hereditary condition, may in itself be a cause of chronic valvular disease in later life, as has been already stated in discussing the causes of chronic endocarditis. It increases the shock of every closure of the aortic valves, and renders necessary more powerful contraction of the left ventricle to drive on the blood, and thus increases the stress on the mitral valve and its tendinous cords. The unremitting strain thus imposed on both aortic and mitral valves is more injurious than the occasional strain due to violent muscular efforts, and tends to set up a chronic inflammatory change which is progressive. Hence, when there is high arterial tension in addition to valvular disease, it will greatly tend to aggravate the mischief already effected, and will contribute a grave addition to the unfavourable elements of prognosis, unless carefully watched and relieved by diet and treatment.

Habits and Mode of Life of the Patient.

These have a very important bearing on the prognosis. Violent efforts or sustained exertion impose a great strain on the valves of the heart; vicissitudes of temperature tax its powers of accommodation to different conditions of circulation, while unfavourable hygienic influences tend to malnutrition and degeneration. The man, therefore, who must labour with his hands, who is exposed to all weathers, whose food is of inferior quality and sometimes insufficient in quantity, who breathes impure air and indulges perhaps in strong drink, who seeks advice only when he can no longer toil, and abandons all precautions as soon as he leaves the hospital, has far less chance of long life than the man who can seek advice early and has adequate means to carry it out. There can be little probability of existing compensation being maintained, or of reparative hypertrophy being established under such circumstances. The first condition of recovery from the effects of disease, the removal of the cause which gives rise to these effects, is wanting. Conversely, protection from adverse influences which have precipitated the access of symptoms, together with rest, warmth, good food, and care, may reverse an apparently hopeless forecast, as is not unfrequently seen in the all but miraculous recoveries which take place in hospitals. In the same way, persistence in habits of eating and drinking, which may lead to valvular disease, directly, by overloading the blood with impurities, which give rise to high arterial tension, or indirectly, through gouty inflammation, will affect the prognosis unfavourably, while obedience to rules of diet carefully laid down and supervised as to their effect will incline the balance to the opposite side.

ANÆMIA.

Anæmia, as is well-known, is very common at the period of adolescence, and especially in girls; and its effects may be most prejudicial. About middle age, essential or pernicious anæmia may complicate heart disease, and in at least two instances which have come under my observation, has been the real cause of death which was attributed to the state of the heart.

But, in addition to primary anæmia, a deterioration of the blood is a common and almost inevitable result of heart disease when this reaches a point at which it begins to affect the circulation. The slow movement of the blood through the systemic capillaries and through the lungs, whether due to deficient *vis a tergo*, as in aortic disease, or to venous stasis in mitral disease, prevents those active changes from taking place by means of which the blood is constantly purified and renewed. Absorption of food, again, will be more or less hindered by the languid movement of the blood in the gastro-intestinal mucous membrane, and by the congestion of the liver, which result from obstruction to the return of blood to the heart.

Anæmia, therefore, existing at a time when prognosis is called for, and especially a tendency to recurring anæmia, is a serious element in the forecast.

Anæmia, however induced, has always a detrimental influence on the heart. It may of itself, aided by the high arterial tension which often accompanies it, give rise to dilatation of the left ventricle and leakage of the mitral valve; it will, therefore, tend to aggravate such dilatation as has already been produced by valvular disease, while it will retard compensatory hypertrophy by impairing the quality of the nutritive material supplied. Anæmia, again, is often attended with palpitation of the heart, and will add to the liability of the valvular disease to this distressing and sometimes dangerous symptom.

When the deficiency of corpuscles has reached a point at which the blood is distinctly watery, the dilute serum becomes thereby increasingly liable to exude from the capillaries into the cellular tissue. Anæmia may thus of itself give rise to œdema, reaching sometimes to a considerable degree, and it will precipitate the occurrence of dropsy when the tendency of the heart disease is in the direction of this complication. In the same way since effusion into the pleural cavity may occur as an effect of anæmia, it may be added to the complications of an affection of the valves and contribute to the danger of the case.

The various points bearing on the prognosis of valvular disease have been enumerated and discussed, but there remain for consideration the circumstances and conditions under which a prognosis may have to be made, and these may be widely different.

1. For instance, we are frequently asked, during convalescence after acute endocarditis—sometimes, indeed, before the attack has subsided—how far the heart is likely to be ultimately affected. This is a question to which no prudent man will give a definite reply. It is impossible at this time to appeal to the changes in the walls and cavities of the heart for guidance, as there has been no time for their development. The cardiac murmurs are not trustworthy guides, as a systolic apex murmur, or sometimes an aortic murmur, which has developed in the course of acute rheumatism will disappear afterwards. If, however, aortic regurgitation is actually established, the prognosis is serious, although no definite opinion can be formed as to the probable rate of progress.

2. When the lesion is one of old standing we may have an individual in apparent health and vigour, scarcely conscious of any inconvenience arising from derangement of the circulation even on exertion, but in whom the discovery of a valvular murmur has been made. There is no modification of the pulse, the murmur accompanies and does not replace the sound with which it is associated, and there is no marked hypertrophy or dilatation of the heart. In such a case the valvular change is slight and unimportant, and of present danger there is none.

With regard to the future of such a patient, everything depends on the question whether the existing state of the valves is an old-standing, permanent condition traceable to a long past attack of rheumatic endocarditis, or is just the beginning of mischief, such as chronic valvulitis, or atheroma, which will go on increasing. In the former case the patient may live to old age and weather all storms of illness and hardship; in the latter, if we take as an illustration the most serious form of disease, aortic regurgitation, he will probably not live more than four or five years, though a definite judgment as to the course and duration of the affection can only be formed after repeated and careful examination at long intervals.

3. In another case, while no symptoms are present, there is hypertrophy or dilatation of the heart, or both. Here the presence of structural changes testifies to the existence of mechanical difficulty due to obstruction or regurgitation, and shows that the valvular lesion is real: there will almost certainly be some corresponding modification of the pulse, and although compensation has been established, the equilibrium may be disturbed by causes, which would have no effect on the normal heart, and once overthrown will not be very easily restored. While, therefore, under favourable circumstances, the health may remain unaffected for many years, an illness of any kind, and

especially an attack of bronchitis, may be attended with dangerous disturbance of the circulation.

In such a case, again, it is of the utmost consequence whether the valvular disease is stationary or progressive: should it be progressive, the prognosis will necessarily be grave, though much will depend on the seat and character of the valvular lesion. The age, sex, general constitutional vigour, and family history must be taken into account, as also the position in life and habits of the patient.

4. In another patient, symptoms of embarrassment of the pulmonary or systemic circulation are present, habitual shortness of breath on slight exertion, violent or irregular action of the heart on slight provocation, which does not readily subside, pain or a feeling of oppression in the præcordial area, incipient œdema about the ankles and perhaps albuminuria with a thick deposit of pink or high-coloured urates in the urine; further evidences of imperfect compensation are also found in the pulse, in the enlargement of the liver and fullness or pulsation of the veins of the neck.

Here danger is never far off and may be imminent, though by suitable precautions it may be guarded against and warded off for years. Speaking generally, there is less probability of prolonged life and comfort after such symptoms have set in, in aortic than in mitral disease. There are more chances of obviating the effects of obstruction by compensatory changes than of making up for failure of *vis a tergo*. As a rule, the earlier in the course of valvular disease symptoms supervene, the more serious is their significance.

5. In another case we are called upon to give a prognosis when grave consequences of a valvular disease have already been developed. The occurrence of these serious symptoms may have taken place in spite of compensatory changes in the shape of hypertrophy of the right or left ventricle or

both, as the case may be, or for the want of them; we must therefore consider whether by opportunity for the restoration or establishment of compensation and by suitable treatment a working equilibrium can be attained. The most important question will be whether the symptoms have been brought on by any temporary or removable cause, such as over-exertion, exposure, anxiety, recent acute illness, pulmonary disease such as bronchitis, anæmia, or debility or the like, or whether there is no recognizable cause for their appearance. In the latter case, the prognosis is far more serious, as some degenerative change in the cardiac walls or organic weakness of the heart or system generally, or possibly some further complication in the shape of adherent pericardium, is to be apprehended. A working man admitted into hospital will almost certainly recover, only however to break down again when he resumes his occupation, and is again exposed to the injurious influences which brought on the symptoms of compensatory failure. A patient in easy circumstances may by care maintain for a long time the *statum quo*, but he will not easily retrace downward steps taken in spite of all favouring conditions. The character of the valvular change loses none of its importance, the only hope of prolonged immunity from further consequences of failing or obstructed circulation will be the absence of any tendency to aggravation of the lesion in the valve. Subject to this the soundness of the patient's organs and tissues and the tenacity of life exhibited by the family history will be elements of great consequence.

6. When we are called upon to form an opinion of the chances of recovery of a patient who is suffering from advanced dropsy, with severe pulmonary congestion and extreme dyspnœa, then indeed the stationary or progressive character of the lesion has no longer any bearing on the immediate issue of the case. If the symptoms have been gradually increasing in severity without any apparent cause,

and the dropsy has crept on from the legs and invaded
the trunk, and the dyspnœa is extreme even while the
patient is at rest in bed, then there is little hope of recovery.
If, on the other hand, the access of severe symptoms is
traceable to over-exertion or a chill or intercurrent pulmonary trouble, such as bronchitis, then there is a hope that
suitable treatment may for a time restore the compensatory
balance, provided that there is evidence of a certain degree
of vigour and force in the cardiac impulse, more especially
in that of the right ventricle. If it is a second attack of
this kind, there will be less chance of recovery, and any
complication, such as kidney disease, will diminish materially this chance. The occurrence of a pulmonary apoplexy
at this stage, or of thrombosis of the veins of the leg, will
be of very serious import, rendering the prognosis as regards
the prolongation of life, even for a short period, almost
hopeless.

CHAPTER VIII.

TREATMENT.

TREATMENT OF VALVULAR DISEASE IN GENERAL—PROPHYLACTIC MEASURES—GENERAL RULES, IN CASES WHERE LESION IS NOT OF SERIOUS EXTENT, AS TO EXERCISE: OERTEL AND SCHOTT TREATMENTS, CLIMATE, CHOICE OF RESIDENCE, DIET, STIMULANTS—TREATMENT OF ANÆMIA AS COMPLICATION—TREATMENT WHERE LESION IS OF MORE SERIOUS NATURE AND HAS GIVEN RISE TO MARKED HYPERTROPHY AND DILATATION OF THE HEART—PRECAUTIONS TO BE TAKEN—SELECTION OF WINTER RESORT—IMPORTANCE OF REST—DIET—STIMULANTS—EMPLOYMENT OF DRUGS—PURGATIVES, THEIR IMPORTANCE—TREATMENT OF VENOUS CONGESTION—VENESECTION—TREATMENT OF THE CONDITION OF ASYSTOLE IN AORTIC DISEASE—DIGITALIS IN AORTIC INCOMPETENCE.

ON THE TREATMENT OF VALVULAR DISEASE IN GENERAL.

THE treatment of valvular disease of the heart has been foreshadowed in the consideration of its prognosis. Whatever influences have been seen to affect unfavourably the condition of the circulation and of the patient, such must be averted or counteracted, and injurious tendencies arising out of the particular lesions of the valves must be combated. The order also in which the therapeutic methods and resources employed for the above ends must be discussed is dictated by the arrangement adopted in the consideration of the prognosis.

One of the most important prophylactic measures is the prevention, as far as possible, of the ill effects on the heart of a recently established valvular lesion. When an attack of endocarditis has subsided, leaving behind it some valvular lesion, the exercise of caution and care will prevent undue cardiac dilatation, which might prove rapidly fatal or cripple the heart seriously in the future.

The patient should be kept in bed, or confined to his room for some weeks, or perhaps even months, according to the degree of severity of the lesion, cases of aortic incompetence requiring the longest period of rest. The object of protracted rest is to allow time for the necessary compensatory hypertrophy of the cardiac walls to take place. In many cases, more especially when pericarditis accompanies the endocarditis, as it frequently does in children, the heart is left dilated and weakened after the subsidence of the attack, so that the risk from premature exertion will be doubly serious, and the period of rest required afterwards will be proportionately longer.

In children, in whom the joint manifestations of rheumatism are usually slight, while the heart is frequently attacked, the heart should be examined from time to time if there is the slightest suspicion of rheumatism, such as fugitive pains in the joints or limbs, with rise of temperature, still more if there are obvious manifestations such as rheumatic nodules; for the onset of pericarditis or endocarditis is often very insidious, and serious mischief may result before it is detected, if the child is allowed to go about as usual.

When the lesion is established, and sufficient rest after the attack of endocarditis has been allowed, the next question to be settled will be, how far the patient may live his ordinary life, or what rules must be laid down for his future conduct, and what precautions should be taken. Everything will of course depend on the nature and the degree of severity of the lesion, and in discussing the treatment the less severe class of cases will be first considered, in which there is no evidence in the shape of marked cardiac dilatation or hypertrophy that the lesion is considerable.

Exercise.—Almost the first question will be as to the rules to be laid down with regard to exercise and exposure to changes of temperature. These will necessarily vary according to the character and seat of the lesion, but not so only—the constitution, strength, habits, and disposition of the patient will also have to be considered. One man is timid and apprehensive; he will scarcely move out of doors lest he should overtax his heart, or eat a sufficient meal for fear of palpitation, or go away for change of air lest he should die away from home. He must be encouraged or even compelled to take exercise. Another is only too ready to ignore the state of his heart, and will run or row or swim or take part in violent games; he must be warned against imprudence, and it may be necessary to forbid such forms of exertion as are liable to be indulged in to excess. In another, an eager temperament and impetuous disposition may reside in a weakly frame, and ordinary duties and an average amount of work may be done with injurious energy and haste. It is not possible, therefore, to draw up definite regulations applicable to all cases. To make the restrictions imposed too severe will in one person directly injure the sufferer's health; in another, will make him unnecessarily depressed and miserable; in a third, will provoke revolt and lead to rash and dangerous violation of all rules. The principle on which recommendations must be based will be to interfere as little as possible with the avocation, habits, and mode of life of the patient, as long as these are not injurious, and especially to allow a maximum of exercise in fresh air compatible with safety.

Nothing can be worse than to debar all patients who are found to have valvular disease from games and vigorous exercise, and to forbid them to go upstairs or to walk uphill, and on no cases do I look back with greater satisfaction than on those, and they have not been few, in which I have liberated boys and girls from such orders. In the class of cases under consideration, supposing a sufficient time to have elapsed after the acute attack in which the valvular affection was established for the necessary compensatory changes to take place, a girl may be allowed to take long walks, to play lawn tennis, to ride, cycle, swim, and dance, and a boy to play cricket and racquets, to hunt, row, box and fence, provided that these exercises are not attended with undue breathlessness and distress, and that they are entered upon gradually and practised with moderation and discretion. On the other hand, football, paper-chases, long house-runs, training for races of any kind, are scarcely permissible.

While discussing the question of exercise the Œrtel and Schott methods may be described.

The **Œrtel Treatment** consists in systematic, graduated muscular exercise carried out at a certain elevation, about two thousand feet above sea-level. The patient is required to walk a certain distance up a gentle ascent each day, the distance and pace being gradually increased. At the same time the diet is carefully regulated, and the amount of fluids ingested is strictly limited. The object of the treatment is: firstly, to stimulate the heart by muscular exercise, carefully adjusted to the capacity of the patient, so as to bring about hypertrophy of its walls; secondly, to diminish the volume of blood in circulation by restricting the amount of water consumed, and increasing the amount eliminated. Œrtel claims that this treatment is successful in cases of fatty heart uncomplicated by disease of the coronary arteries, in cardiac dilatation, and in valvular disease, even when compensation has broken down, and dropsy with other evidence of venous congestion is present.

In cases of valvular disease, in which compensation has completely given way, this treatment is certainly not advisable, nor in many instances would it be possible; but where compensation has been established after a recent valvular lesion, or has been restored by rest and suitable treatment, or where it is maintained with some difficulty, gentle climbing exercise in fresh and pure and somewhat rarefied air will certainly do more to develop further compensatory hypertrophy of the heart, than mere walking on the level, which has not the same beneficial effect on the circulation, and where the air is not so pure or invigorating.

It is more especially in cases of fatty heart without fatty degeneration of the cardiac muscle, arising from overeating and drinking and insufficient exercise, that this treatment by dieting and systematic muscular exercise may be of real service. At the commencement of the treatment, great caution should be observed as to the nature and amount of exercise. The patient should only be allowed to walk a certain distance up a gentle slope, and each day the distance may be gradually increased. By this means, and by the limitation of the fluids ingested, the superfluous adipose tissue is gradually got rid of, and the tone of the heart muscle is restored. Not infrequently this treatment is employed as a sequel to the Schott methods.

The Schott Treatment.—The treatment by baths and exercises by the method of Schott, of Nauheim, may be of service in suitable cases. The waters of Nauheim are remarkably rich in free carbonic acid gas, as well as in mineral constituents, the chief of which are chloride of sodium, and chloride of calcium, and carbonate of iron. At the beginning of the treatment the baths should contain about 1 per cent. of chloride of sodium, and 1 per 1000 of calcium chloride, and should be free from carbonic acid gas. The bath at first should last from six to eight minutes, and should be of the temperature of

92° to 95° Fahr. As time goes on the proportion of solids in the bath should be increased, also the duration of the bath, while the temperature is gradually lowered. Eventually, baths containing free carbonic acid gas, and about 3 per cent. of sodium chloride and 3 per 1000 of calcium chloride, may be taken. Baths can be artificially prepared in imitation of those at Nauheim, the essential ingredients being the chlorides of sodium and calcium and the free carbonic acid gas.

The exercises consist of a series of simple movements of each limb and of the trunk made against slight resistance, so that every muscle of the body, as far as possible, is in turn brought into play. The movements should be made slowly and systematically, and a short interval of rest should be interposed between each; they should be stopped if the patient experiences any distress in breathing or discomfort, but may be proceeded with again as soon as he is rested. The movements consist of flexions, extensions, adductions, abductions, and rotations of each limb in turn, and of flexion, extension, and rotation of the trunk.

The effects produced on the circulation and the heart, whether by the baths or exercises are similar; the pulse frequency is diminished, its volume and force are increased, and the area of cardiac dulness in cases of cardiac dilatation is diminished, while the apex beat recedes and comes nearer to the normal position.

Such are the immediate results which appear to indicate an improvement in the contractile power of the heart and a reduction of its dilatation; but these are not permanent. A long course of baths and exercises is therefore necessary. The ultimate object of the treatment is that the improvement in the condition of the heart and pulse thus manifested should become more pronounced and of longer duration each time and ultimately become permanent.

Schott's theory is that, as a result of each bath or

series of exercises, the heart is stimulated by a reflex process, so that its contractions become more complete and forcible, and that as a result of this frequent stimulation the heart muscle undergoes hypertrophy, and so becomes competent to cope with the extra work thrown on it by a valvular lesion, or recovers from its atonic condition. It seems more probable that the baths or resisted movements give rise to a physiological dilatation of the capillaries in the skin or muscles respectively, so that the resistance to the onward flow of blood is lessened, and the left ventricle thus relieved is able to complete its systole. At the same time, from the moderate and gentle character of the exercises, compression of the veins, such as occurs in severe muscular exertion, driving on the blood to the right ventricle and causing dyspnœa, does not take place. There is thus a transfer of blood from the venous to the arterial system, which is the reverse of the tendency in most forms of heart disease. The chief objection to this theory is the slowing of the pulse that occurs in the bath or during the exercises, as diminished peripheral resistance would rather tend to accelerate than slow the pulse. Possibly the slowing of the pulse is attributable to reflex stimulation of the vagus.

This treatment by baths and exercises has been practised for many years at Nauheim, but has only lately been introduced in England, and many successful results and cures have been claimed from its employment. It cannot of course cure valvular disease, in the sense of causing the vegetations or deformities of the valves to disappear, but it may give relief and greatly modify the symptoms in suitable cases. It must not, however, be thought that this treatment is applicable to and infallible in all varieties and conditions of morbus cardis, or that it is to be a substitute for all other forms of treatment.

In cases of cardiac dilatation from loss of tone of the heart muscle after influenza or some depressing disease, it may be of great service, and effect a cure where drugs and other treatment have failed: in many cases of functional and neurotic heart disease, which are very common and are difficult to deal with, it may also give satisfactory results. In valvular disease it is of course unnecessary, when compensation is established and no symptoms are present: when compensation has completely broken down, it is not advisable, as rest in bed and suitable treatment of other means will be more efficacious. In cases of mitral disease, more especially mitral stenosis, when compensation is just maintained with difficulty, and when the degree of stenosis is such, that increased contractile power of the right ventricle induced by digitalis would be useless or harmful, it may be of great service.

In aortic disease it is not advisable, owing to the risk of syncopal attacks, though when compensation is breaking down and mitral symptoms are present, it may sometimes yield good results.

In adherent pericardium with threatened compensatory failure, it may be of service. This treatment may not yield such good results in England as at Nauheim, as, apart from the effects of the natural mineral baths there, and the exercises, the change of air and scene, the quiet uneventful life, the early hours and regular meals, together with freedom from all excitement and worry, will greatly contribute to the success of the other remedies.

Exposure to Change of Temperature.—As regards exposure to vicissitudes of temperature and to changes of weather, we have to bear in mind that the subjects of valvular disease have, in most cases, already manifested a susceptibility to the effects of cold and wet, or an inherited predisposition to rheumatism. One of the things, therefore, most to be feared and guarded against, is another attack of rheumatism. Flannel or woollen underclothing of some kind should be worn next to the skin winter and summer, and standing or sitting with wet feet or in damp clothes, or lying down on the grass after games, involving the risk of getting chilled after perspiration, must be forbidden. But we must not go to the other extreme and cultivate an undue susceptibility to cold by excessive care. The patient need not be kept indoors by rain or cold, or forbidden to get into a perspiration. Excessive precautions defeat the object for which they were enjoined, and it is easy to reduce the power of resistance to changes of temperature until the slightest exposure is attended with risk. While, however, the primary aim ought to be to maintain and confirm such hardihood as the constitution possesses or is capable of, there are cases in which safety consists in flight from too severe a climate, or from a situation or soil conducive to rheumatism. There is full opportunity for the exercise of judgment in deciding how much may be dared and how soon the patient must yield.

Climate.—Climate may do much to influence the course of heart disease. On *a priori* grounds, we should say that a mild, dry, bracing and equable climate, in which the patient could have a maximum of exercise in the open air, would be the best, but, independently of the question where these desiderata are to be found, all such general statements have to be qualified to meet the idiosyncrasies of individuals; the great majority of our patients, moreover, are tied down by circumstances to a particular spot, and the most important practical problem usually is how to make the best of a given neighbourhood.

Choice of Residence.—If the choice of a residence is open, we should direct the patient to seek a gravel or sandy soil at a moderate elevation, where the rainfall is below and the sunshine above the average, and the water not hard—conditions best realized, in England, in Kent, Surrey, and Sussex. The exposure of the house should be to the south, and there should be protection from the north and east.

The immediately surrounding country should not be too hilly, and especially the house itself should not be on the top of a steep hill, making every walk necessarily end in an ascent. There would be danger, some day or other, of dilatation or other ill effect from the exertion after a longer or more tiring walk than usual or when out of sorts. It must always be borne in mind that the good results obtainable from the best climate may be neutralized by faults of detail in the placing or construction of a house.

Diet.—About food little need be said. Excess should be avoided, but, subject to this condition, the diet may be liberal and varied. It is important, however, that there should be a due proportion of farinaceous and vegetable articles of diet; when the food is highly nitrogenized, as when it consists largely of meat, imperfectly oxidized waste accumulates in the blood, and this is a great cause of resistance in the capillary circulation, which constitutes a serious addition to the work imposed upon the heart, and puts a continued strain upon the compensation by which it adjusts itself to the imperfect state of the valves. This recommendation is specially important in the case of constitutions disposed to gout, since the valves may be further damaged by gouty inflammation. Habitual excess of food beyond the requirements of the system will have the same effect.

The food again should be divided into three fairly equal meals, and not taken in excessive quantity at dinner, or at breakfast and dinner. If the nourishment for the twenty-four hours is consumed at one huge repast, the blood is drawn upon for the consumption of the tissues and for the supply of the secretions in the long interval, and its volume being reduced, the vessels are depleted; the products of digestion are then rapidly absorbed, the amount of blood is increased, and the vascular system is filled and perhaps overcharged.

When its valvular apparatus is unsound, or its structure is impaired, the heart does not easily adjust itself to such extremes, and, if it so far effects this that no particular discomfort is experienced, the increased work thrown upon a weak organ cannot fail to be injurious in the long run. A very large meal, again, must distend the stomach, and the diaphragm may be pushed up and hindered in its action so as to embarrass the heart directly by pressure, and indirectly by interfering with respiration. A dilated stomach by upward pressure may so embarrass a heart which is diseased as to give rise to irregularity of rhythm, anginoid pains and even syncopal attacks, one of which may prove fatal.

Stimulants.—Strict moderation must be observed in the matter of alcoholic drinks; in comparatively few cases are they necessary, and if taken they should be taken only as part of a substantial meal. Their effects as excitants of the heart may, to some extent, be neutralized by the relaxation of the peripheral vessels which they induce, but their general tendency is to interfere with due metabolism and elimination, and to bring about degeneration of structure.

Unduly high arterial tension, from whatever cause, must be combated. Its injurious effects have been already pointed out. The regulations as to diet and drink have for one of their objects the prevention of high blood-pressure. When this condition exists from inherited tendency, or from gout or renal disease, it must be kept down within safe limits by suitable eliminants.

Regulation of Bowels.—It is always important to take measures against constipation. Accumulation of fæcal matters in the large intestine, with the associated flatulent distension, will more or less embarrass the heart, both by direct pressure upwards of the diaphragm and indirectly by interference with respiratory movements. Palpitation, again, is a frequent result of constipation, and both the effort required to unload the bowel and the different pressure on the abdominal veins before and after a large evacuation put stress upon the heart. A further ill-result is the retention of toxic matters in the blood which provoke resistance in the capillaries and tend to the production of high tension.

Anæmia.—While everything is done to maintain the general health, special precautions must be taken to guard against anæmia. We have seen that it predisposes indirectly to valvular lesions, and that it may itself give rise to dilatation of the left ventricle, which is the most serious aggravation of valvular disease. Valvular disease, moreover, tends to deteriorate the blood in so far as it interferes with its free movement through the glands and tissues generally, and to produce anæmia. It does not follow that we are to be always giving iron or other reconstituent tonics, but it is a valid reason for careful choice of residence and for frequent change of air, and for attention to and treatment of the earliest indications of anæmia.

Treatment in Cases of severe Valvular Lesion.

When the presence of dilatation and hypertrophy indicate that the damage to the valve has been such as to interfere perceptibly with transmission of the blood through the heart, while the principles with regard to exercise remain the same, some modification in their application will be necessitated. Each form of valvular disease gives rise to a certain kind and degree of interference with the efficient pumping of the blood through the heart, lungs, and system, which tends to the production of results injurious to health, and in the long run dangerous to life. We must not wait for the recognition of such tendencies till they are forced upon our attention by the appearance of symptoms; they can always be foreseen, and they may often be prevented. Again, it must be borne in mind that when valvular disease of any importance, as indicated by changes in the walls and cavities, exists, the heart has to some extent lost the power of responding to sudden calls for variations in the rate or force of the circulation; more than this, its power of recovering itself after disturbance is impaired. The compensation effected by hypertrophy is efficient only for ordinary purposes, or within certain limits more or less restricted. When the muscular pressure on the veins all over the body, which attends vigorous exercise, brings the blood in increased quantity to the right cavities of the heart, it cannot be sent on against the resistance in the pulmonary circulation, which results, as we have seen, from most forms of valvular disease. Even in health there is a period of breathlessness before the right ventricle succeeds in driving the blood through the lungs as fast as it arrives from the system, when we begin to run, and in disease, this breathlessness is more readily provoked and is easily exaggerated to painful dyspnœa; the distended condition of the right ventricle and auricle, which is induced by exertion in health, and which acts as a reservoir for the blood till it can be delivered to the lungs by the increased action of the heart, may already exist in disease, so that the provision for the emergency is already exhausted, and there is no margin left for severe exertion.

While, therefore, we avoid unnecessary and injurious restrictions on exercise, sudden and violent exertions must be forbidden, and it must be understood that anything which gives rise to painful breathlessness is injurious. A patient, however, will often, by beginning gradually, arrive at a rate of walking which, attempted at first would have been impossible, and may eventually, by the exercise of similar caution, mount with ease an incline which would otherwise have brought him to a standstill. So long as the equilibrium of the circulation is not disturbed, there is no particular danger in going uphill. The danger arises from the fact that on coming to an ascent, the tendency is to maintain the same pace as on level ground until we are checked by shortness of breath. The sufferer from heart disease cannot afford to do this. What would be a mere fugitive inconvenience to a man in health may be a risk and injury to him. If, however, he will exercise a little foresight and slacken his pace immediately on coming to an incline, he may afterwards gradually increase it again within certain limits and so climb hills. The same conclusions apply to stairs. So long as the subject of a valvular affection can go upstairs quite comfortably, there is no objection to his doing so; and if by ascending them quietly he avoids breathlessness, this need not be forbidden even when symptoms are already manifest. Sometimes he can go up backwards without distress, when to take them step by step in the ordinary way is difficult. But there comes a time when stairs must be avoided as far as possible, and when the patient must be carried up or must live on one floor. This will especially be the case when not mere breathlessness but faintness is produced by slight exertion, and the patient feels giddy or has dimness or temporary loss of vision, or weakness and trembling of the knees after going uphill or upstairs. It is in aortic disease that such symptoms are most likely to be experienced, but they may occur whenever the left ventricle is greatly dilated or is weak from any cause.

Age.—In applying any rules for the management of valvular disease of the heart, a great difference will be made between the young and those who have reached or passed middle age. After a certain period of life, varying greatly according to constitution and habits, there is a liability to dilatation of the heart on exertion, and this condition, attended at once by symptoms and leading to a speedy fatal issue, is often brought about by an imprudent effort independently of any pre-existing affection of the valves.

The risk of such an event is indefinitely increased when dilatation, attendant on valvular lesion, is already present; or when the tendency thereto has only been neutralized by compensatory hypertrophy. The ventricle is tried by chronic overwork, and the nutrition of the increased amount of muscular tissue in the walls is maintained with difficulty, the heart is consequently more ready to break down under stress. We may, therefore, allow a boy to play cricket, or the young of either sex to play lawn tennis, to boat, or even swim in moderation, provided always that no distress of breathing or tendency to syncope is induced, while a corresponding amount of exertion would be altogether forbidden at middle age.

Under no circumstances, and at no age must fatigue be carried to the point of exhaustion. There is less chance of recovery from the effects of overtaxed endurance than from those of a brief violent effort.

Selection of Wintering or Holiday Place.—In deciding upon a place for temporary change of air for a patient suffering from valvular disease, we must be guided very much by the previous experience of the patient. In some cases a trip by sea will be of the greatest service, but this could be recommended only to good sailors. A winter in Egypt, Algiers, or Rome, may be good in others, or if this is not practicable a few months on the south coast of England, at places where exercise can be taken without much up and downhill. In summer the seaside is most generally useful, but some people tell us it does not suit them, that it makes them bilious. This is often a mere temporary effect, which quickly passes off, or it may be a result of the constipation, which is common. In all cases constipation must be provided against.

When change of air is recommended, great caution must be exercised in sending patients suffering from heart disease to any considerable height, such as 5000 or 6000 feet. The effects of the reduction of atmospheric pressure cannot be foreseen; palpitation is often set up which is absolutely intractable without removal to a lower level, and this is sometimes difficult. One reason for this may possibly be the expansion of the gases in the stomach and intestines, permitted by the diminution in the atmospheric pressure at high altitudes. Every year people arrive in the Engadine with unsuspected heart disease, which is revealed or perhaps developed by palpitation and dyspnœa, rendering a hasty retreat imperative. On the other hand, in a case in which the absence of dilatation and hypertrophy shows the valvular lesion to be slight and the patient is capable of all ordinary forms of exercise without suffering, high altitudes will usually give rise to no inconvenience or danger.

Moderate heights, say 2000 or 3000 feet, are usually well borne, even when there is decided valvular disease, and graduated exercise at such elevations is systematically employed as a means of improving the tone and vigour of the muscular walls of the heart, and of removing fatty deposit which may have taken place upon its surface and in its substance.

Complete Rest in Bed.—Rest is, in many cases, an important part of the treatment, and when the state of the patient demands it, rest in bed is of the greatest service; but one of the questions calling for the greatest exercise of judgment is to decide when absolute rest is necessary. There are circumstances in which a month or six weeks in bed will prolong life for as many years. On the other hand, the disturbance of compensation and the supervention of symptoms often date from some slight injury which has confined the patient to his couch or to the house for a, month or two. Even when severe symptoms are present to insist upon confinement indoors is sometimes to add to the suffering of the last few months of life without adding to its duration. When the onset or exacerbation of symptoms is distinctly attributable to work persisted in from necessity, as in the case of most hospital patients, or from courage and defiance of pain, as we sometimes see; or when the aggravation is due to imprudent exertion or exposure, or to some intercurrent pulmonary complication or other illness, there can be no hesitation in ordering rest. The doubt arises in cases in which the effects of valvular disease are gradually creeping on, and the increasing breathlessness and evening œdema are the direct result of the obstruction to the circulation and of failing compensation. Under such circumstances it is difficult to decide when to interfere. The patient will say he is worse in bed than up, that he cannot lie down, but has to be propped up by pillows; that he cannot and dare not sleep, and that if he drop off into a dose he wakes up in indescribable fright and distress. His attendants, it is true, may tell us that he has had more sleep than he supposes, but all the same the nights are long and miserable, and it is no light matter to condemn a sufferer, who has looked and longed for morning, to a couch which, in his experience, is associated with his worst moments. Not uncommonly, however, after sleepless tossing till 2, 3, or 4 a.m., with inability to lie down, quiet sleep may come, and the sufferer may slip down into a comfortable position in which he gets real repose, and sometimes after twenty-four or forty-eight hours in bed, relief is experienced, which reconciles the patient to the confinement. We must be guided by results and by our knowledge of the patient.

A similar difficulty frequently confronts us in advanced stages of heart disease, especially when attended with dropsy. The patient implores us to be allowed to sit up, and in paroxysms of dyspnœa he is compelled to do so and to throw his legs out of bed and let them hang down. It is an unequivocal dictate of experience, not easily accounted for by theory, that the heart in disease is often relieved by an upright posture of the body and a dependent position of the legs. Such is the case, not merely because in dropsy the swollen abdomen and thighs make it impossible to sit well forward in bed when the legs are raised, but because in some way the position facilitates the action of the heart, it may be by taking off the pressure of the abdominal viscera from the diaphragm, or by allowing blood to gravitate from the right auricle into the vena cava inferior and abdominal veins, or it may be in consequence of a physiological diminution of the arterial tension which, according to Dr. Oliver, attends the erect position. There are few medical men of large experience who have not seen instances in which the sufferer has not gone to bed for months, but has slept all this time in an armchair with some support contrived for the head. On the other hand, our results are better when the patient can be kept in bed; there is a better chance of removal of dropsical effusion and of recovery generally in the recumbent position. While, therefore, we do not carry too far our resistance to a patient's entreaties to be allowed to sit up in a chair for a part or the whole of a day, recognizing, also, that a few days or weeks of life may be dearly purchased if it is at the expense of increased suffering, it is our duty sometimes, and especially when there is a chance of recovery from the existing complications, to exercise firmness in keeping the patient in bed in spite of great temporary distress.

There is less uncertainty with regard to exposure to cold, though in an advanced stage the patient often complains of subjective heat, and throws off his coverings, or insists on the window being opened on the coldest day to satisfy his want of air. External cold contracts the arterioles of the surface, and increases the resistance in the systemic circulation, and it should be prevented from reaching the sufferer. The open window, however, may be permitted, on condition that the patient is efficiently protected. The play of cool, fresh air on the face, and a deep draught of it into the lungs, are indescribably refreshing. The cold and livid lower limbs in aggravated dropsy appear to be almost insensible to changes of temperature, but they should none the less be carefully covered.

Diet.—Little need be said about nourishment. The patient may have whatever he can eat and digest, but his appetite will leave deficiencies, which must be supplied by liquid food of different kinds, all forms of which may have to be laid under contribution.

Stimulants.—Alcoholic stimulants afford invaluable help and will be required more or less in every case of any severity. At first a small quantity of any wine which suits the patient may be taken with food, and frequently a little spirit in hot water at night will help to procure sleep. Ultimately stimulants, especially spirits, may have to be given very freely; but great caution must be exercised in the early periods, since the struggle against the encroachments of the malady is often very long and trying, and if the good effects of alcohol are exhausted rapidly, the patient is left without resource when the time of greatest need arrives. As to the actual amount required, I should consider about ten ounces of brandy per diem the maximum likely to be useful in the most urgent cases. This should be reached very gradually from two or three ounces, and whenever an emergency has led to an increase of the allowance, it should be reduced when the occasion has passed. We must, above all, be careful not to be misled by the patient's demands. He is conscious of relief from the stimulant, and not unnaturally asks for it whenever the oppression or depression becomes severe; but it is easy to pass the limits of usefulness, and to produce a feeling of depression which is only reaction from an excess of alcohol.

In larger quantities than ten ounces the effects become

uncertain, and I have often seen improvement from diminution of a dose which seemed to be imperatively required.

When symptoms have arisen requiring the employment of medicinal treatment, the first question to be considered is whether the heart can be relieved in any degree of work to which it is no longer equal. It is very rarely that this is not the case, and not uncommonly a lessening of stress upon the heart is sufficient of itself to restore the circulatory equilibrium. A diminution in the volume of the blood by eliminants of various kinds, removal of portal congestion and of distension of the abdominal veins by purgatives, which will relieve the right side of the heart and lower the arterial tension, are among the measures most generally useful and most commonly required.

The fatal result is reached through various secondary consequences, the prevention or removal of which postpones the final issue, and not only does this, but also relieves suffering. These consequences, therefore, must be studied, and it will be necessary to revert to the modes of death from valvular disease of the heart and to consider the tendencies thereto which we have to counteract. Leaving out of consideration embolism, whether systemic or pulmonary, which will best be averted by the prevention of stagnation of blood in the heart, heart disease tends to the arrest of the circulation in two different ways—by failure of propulsion through the arteries and by damming back in the veins. It is not to be understood that one or other of these tendencies is alone in operation in any given case; the rule is that both are present, as has been already seen. While, however, in most cases of valvular disease which have proceeded so far as to give rise to serious symptoms, the double effect on the circulation—the damming back in the veins and the imperfect propulsion through the arteries —is recognizable, one or other will be a primary tendency, and will predominate, and the treatment must be directed to the rectification of the tendency which is most concerned in placing life in danger. By far the more common is venous stasis, since it not only follows directly from both forms of mitral disease, but may also be a secondary result of aortic disease, and it gives rise to a characteristic train of phenomena. Inadequate arterial supply is of itself more likely to cause sudden death than to produce symptoms.

With venous obstruction the liver will be enlarged and greatly congested, perhaps pulsating, and one of the first objects of treatment is the relief of this engorgement of the liver. Thereby relief will also be afforded to the nausea and sickness dependent on the congested state of the gastro-intestinal mucous membrane, and also to the over-distended right side of the heart, which is constantly receiving from the liver, which acts as a kind of reservoir, more blood than the right ventricle can transmit through the lungs. The means to be employed for the purpose are chiefly aperients, and all purgatives will have the desired effect in a greater or less degree; but it is not a matter of indifference what drugs we employ. The best results are undoubtedly to be obtained, according to my experience, from purgatives, in which calomel or other mercurial preparation is a constituent, such as calomel and compound jalap powder, calomel, blue pill, or grey powder, with colocynth and hyoscyamus, followed or not by salines. Hydragogue cathartics of greater violence may be necessary in some cases, but the effect on the liver and heart is not proportional to the degree of purgation, and the relief of dropsy is not due simply to the amount of liquid carried off by the intestinal surface, but is frequently the effect rather of the diuresis which follows improvement in the circulation. Digitalis is often useless, and appears only to add to the embarrassment of the heart and to produce sickness, until the way has been cleared for its operation by a mercurial purge, and where its good effects on the heart seem to be expended, a fresh start will often follow a calomel and colocynth pill.

A troublesome, watery diarrhœa, which is not unfrequently present, is no contra-indication for purgatives; but, on the other hand, constitutes a distinct call for a decided aperient. It is due to the passive congestion of the gastro-intestinal mucous membrane which results from the obstruction in the portal system, and is relieved by free secretion from the mucous surface and from the liver.

But the venous obstruction may reach a point which it is out of the power of purgatives to affect. The lower edge of the liver is at or near the level of the umbilicus (or this organ may be prevented from swelling by cirrhosis), the right cavities of the heart are almost paralysed by over-distension, which, with regurgitation through the tricuspid orifice, reduces the transmission of blood through the lungs to a minimum, and the left ventricle receiving little blood has little to forward into the arterial system. The pulse is weak, small, irregular, both from the irregular action of the heart usually present, and from some of its beats not reaching the wrist. Unless the circulation is to come to a standstill the right side of the heart must be promptly relieved of the over-distension which is the immediate cause of the threatened arrest, and this can be most quickly and effectually done by venesection. The withdrawal of a few ounces of blood (8-16) so far reduces the pressure in the right auricle and ventricle when the latter is checked at the very beginning of its systole by resistance which it is unable to overcome, that it regains command over its contents, and is once more able to drive the blood through the lungs. Nothing can be more striking or satisfactory than the effect of bleeding from the arm under such conditions. The face may be livid and bedewed with cold sweat, the extremities blue and cold, but warmth and colour will return, the dyspnœa will be relieved, and the pulse will improve. The condition of the right ventricle is of critical importance when venesection appears to be required. If it is weak and degenerating, and unable to take advantage of the relief afforded by the withdrawal of blood, the desired result does not follow. We ought, therefore, for bleeding to be successful, to have a powerful right ventricle impulse heaving the left costal cartilages, and perhaps the lower end of the sternum itself, and felt below the costal margin.

In some cases, where the circulation is so nearly at a standstill that blood will not flow from the vein when opened, a hypodermic injection of ether or strychnia in the præcordial region may cause the blood to start flowing and save the patient from impending death.

Very frequently, especially in hospital patients, the desired relief may be obtained by leeches, and I have usually selected the region of the liver for their application, not of course with the idea of taking blood from this organ, but because there is, as a rule, local pain here which is relieved: six or eight may be applied at a time. The abstraction of blood will be followed up by a mercurial purge and digitalis. Brandy and other stimulants may be given at the same time; there is nothing inconsistent in helping the oppressed organ at the same time that it is being relieved from the special difficulty with which it has to contend.

When, as in aortic incompetence, imperfect propulsion of blood by the left ventricle—asystole, as it may be called—is the cause likely to lead to arrest of the circulation, there is no such conspicuous train of symptoms. There may be præcordial pain, dyspnœa, syncopal attacks, etc.; but since a momentary failure of the circulation in the vital nerve centres is fatal, sudden death may occur before the warnings of danger have arrested attention.

Such warnings are sudden attacks of faintness and dimness of vision, giddiness, anginoid pains, sudden weakness and trembling of the knees.

In the condition of asystole, measures suitable for the relief of venous stagnation would be fatal. We cannot resort to bleeding; purgation must be employed with caution. The treatment must rather be directed to stimulation of the failing heart, by drugs such as strychnine, ammonia, and ether, together with good, easily digested, nourishing food, and alcohol regularly administered in small quantities. Vascular dilators, such as nitro-glycerine and nitrite of amyl and sodium nitrite, or perhaps erythrol, or mannitol nitrate, the effects of which are more lasting according to Bradbury, may be of great service, more especially in cases where anginoid pains are a prominent symptom.

Digitalis may sometimes be of great service, but its effects are not constant, and may be unfavourable, or may become so after a period of marked benefit. The reasons for the apparently uncertain and inconstant results of digitalis in aortic incompetence will be discussed in the next chapter.

CHAPTER IX.

ABUSE OF DIGITALIS—SUBSTITUTES FOR DIGITALIS—THE GROUP OF CARDIAC TONICS OF THE DIGITALIS TYPE—THEIR PHYSIOLOGICAL ACTION: THERAPEUTIC EFFECTS—USE OF DIGITALIS IN AORTIC STENOSIS, IN MITRAL INCOMPETENCE, IN MITRAL STENOSIS.

Use of Digitalis.

It is too commonly taken for granted that the existence of valvular disease constitutes an immediate indication for the administration of digitalis. But to make the discovery of a murmur the signal for giving digitalis is fatal to anything like precision in treatment, and may deprive the sufferer of the advantage to be derived from this remedy when it is really needed. The special indications for its use are frequency, weakness, and irregularity of pulse, and œdema of the extremities, with scanty, turbid, concentrated urine. When these are absent, it is rarely of service; but even when these symptoms begin to show themselves gradually or occasionally on slight provocation, it will be well to combat them at first with strychnine, iron, quinine, and general tonics, rather than resort at once to digitalis, the salts of potash and any suitable vegetable diuretic being employed to promote secretion of urine. When the use of digitalis is called for, the most trustworthy evidence of its beneficial effects will be increase in the secretion of urine, with an improvement in the tone and vigour of the pulse, as well as a more regular and less hurried action of the heart. When there is no response in the form of diuresis, the pulse and general symptoms must be carefully watched lest harmful effects should arise.

Digitalis may be given in combination with nux vomica or strychnia, or with pil. hydrarg. and squill, or with salts of ammonium, or iron, according to circumstances, which cannot be minutely laid down. In most cases it will be advisable to give a mercurial purge before its administration, and to repeat this from time to time.

In cases where there is high arterial tension, the mercurial purgative is especially important, and it may be well to give with the digitalis spiritus ætheris nitrosi, or some vaso-dilator to counteract in some measure the tonic effects of the digitalis on the arterioles and capillaries. As alternatives to digitalis, strophanthus, convallaria, caffein, spartein, cactein have been advocated.

Convallaria has been extensively tried, but either its effects or its preparation is uncertain, and it does not appear to be of great service.

Strophanthus may be a most useful alternative when digitalis produces sickness, and may even succeed where digitalis has failed. It is claimed for strophanthus that its tonic action is mainly confined to the heart, and that it does not cause contraction of the arterioles and thus increase the peripheral resistance in the same way as digitalis.

Caffein has been said to increase the solids in the urine, which would make it a useful complement to digitalis, which promotes a flow of water. In my hands it has often been a most useful accessory to digitalis, diuresis and general improvement setting in promptly on the administration of 5 gr. doses of citrate of caffein three times a day in addition to digitalis, which had failed to produce any decided effect without it. I have not, however, found that caffein alone is an efficient substitute for digitalis.

These drugs, which contribute the most important group of cardiac tonics, possess special interest, and merit a separate discussion, since they not only render important service in the treatment of disease, but furnish one of the best illustrations of the relation between physiological and therapeutic action. Digitalis, the best known and longest employed, as well as the most important of them, will be taken as the general representative of the class. Their physiological action will first be briefly mentioned.

Their Physiological Action.

The physiological action is a stimulation of the muscular fibres of the entire cardio-vascular system, giving rise, on the one hand, to more deliberate and powerful action of the heart, and on the other to tonic contraction of the arterioles and capillaries. After death from the poisonous effect of digitalis, the arterioles are narrowed to an impervious thread, and the ventricles of the heart are found firmly contracted upon themselves, and empty. The drug appears to act directly upon the muscular structures, and not through an intermediate influence upon nerves; we have not, therefore, in considering the effects upon the heart, to discuss the question whether they are produced by inhibition of the sympathetic or stimulation of the vagus.

Their Therapeutic Action.

There are various ways in which the physiological action thus briefly sketched may come to the aid of the circulation when the transit of blood through the heart is hampered by valvular or other disease. Besides the more complete expulsion of their contents by the energetic contraction of the ventricles, which will help to fill the arterial side of the circulatory system, there will be improved suction action during diastole, which will tend to withdraw from the veins the blood which has been dammed back and remains stagnating in the liver and abdominal venous plexuses. Another effect is that not only are the force and effectiveness of the systole increased, but the general vigour of the heart is renewed, through the increased physiological rest resulting from the relative prolongation of the diastolic period, which gives opportunity for nutritive repair of the cardiac muscular fibres, and for the reaccumulation of energy expended in the systole.

The primary effect of the tonic contraction of the arterioles will be to increase the resistance in the peripheral circulation, thus throwing more work upon the heart; and it is conceivable that the arterio-capillary contraction may, under certain conditions, more than neutralize the increased force of the ventricular systole, as, for example, when the cardiac muscular fibres have undergone serious degeneration, and, as a matter of observation, this is found sometimes to be the case. For the most part, however, the disease, and especially valvular disease, which has given rise to the necessity for cardiac tonics, will have brought about considerable hypertrophy of the muscular walls of the heart without any corresponding hypertrophy of the muscular walls of the vessels. The balance of advantage, therefore, when the contractile energy of both heart and vessels is increased, is largely on the side of the heart, its muscular fibres having greatly increased in number and size.

It might seem, again, that the contraction of the arterioles would more or less intercept the *vis a tergo* in the veins which is already lacking, as is often manifested by the presence of œdema. If, however, we bear in mind that the contraction affects not only the arterioles but the capillaries also, it will be evident that the narrowing of these channels will give rise to increased rapidity of the current of blood within them which will carry to the venules the propulsive force communicated by the heart better than a sluggish and irregular movement through a network of flaccid and dilated and bulging capillaries. This more rapid onward flow of the blood will again favour the taking up of fluid effused into the inter-cellular spaces.

It would seem that the effects just enumerated ought to be of equal service in all forms of heart disease, excepting structural degeneration of the walls, and certainly in all forms of valular disease. As a matter of observation, however, such is not found to be the case. In one valvular disease, mitral incompetence, all observers agree that digitalis is of the greatest possible service; in another, mitral stenosis, there is almost equal concurrence of opinion that this remedy is not of the same benefit, and, indeed, that it is capable of doing harm and of aggravating the bad effects of the disease. In aortic incompetence opinions are divided, some maintaining that the cardiac tonics in general and digitalis in particular are injurious, others that they are helpful. The same may be said of aortic stenosis.

It appears to me that an explanation of this difference, if it can be arrived at, will make our comprehension of the beneficial effect more clear, and render our employment of these remedies more precise.

The question of the effects of digitalis is often argued on theoretical grounds; but it must be pointed out that it is upon experience and not upon theoretical considerations that the conclusion just stated as to the difference in its remedial influence is based. Long before the physiological action of digitalis was ascertained, it had been noted that this remedy was not always beneficial in its action, and was sometimes obviously injurious, and much was said by old writers as to intolerance of the drug and especially as to its cumulative effects. When mitral stenosis was not distinguished from incompetence of this valve, the varying effects of digitalis must have been most perplexing, and, indeed, incomprehensible.

In aortic regurgitation failure of compensation is manifested in two distinct ways, and there are two different modes of death. In one, the effect is defective propulsion of blood into the arterial system, manifested by faintness, giddiness, and sudden weakness of the legs, sometimes by anginoid pain; death is by syncope; in the other, there is obstructive backworking through the lungs and right heart, giving rise to venous obstruction and dropsy, exactly as in mitral insufficiency. There are, in effect, aortic physical signs with mitral symptoms.

We have in this, it appears to me, an explanation of the different views as to the influence of digitalis in aortic insufficiency. When the tendency indicated by the symptoms is defective propulsion with failure of arterial blood supply to the brain, the effects are uncertain and even doubtful. While sometimes apparently beneficial for a while, a frequent result is the production of irregularity of the pulse with aggravation of the symptoms, and occasionally of vomiting attended with rapidly increasing weakness of the heart's action. Cases are met with, indeed, in which there is grave reason to suspect that it has precipitated a sudden fatal termination.

When, on the other hand, the symptoms are due to secondary dilatation of the left ventricle not adequately neutralized by hypertrophy, with or without mitral regurgitation, and to the effects of this upon the pulmonary circulation and right ventricle, we have exactly the same opportunity for the beneficial influence of digitalis in reinforcing the right side of the heart, and the same favourable results as in mitral regurgitation. The effects, indeed, are sometimes much more striking, and the removal of dropsical effusion more rapid. I have observed, however, that not uncommonly patients suffering from serious aortic insufficiency, after recovering from the mitral symptoms, die suddenly from failure of the left ventricle, and this whether the digitalis has been continued or left off, and sometimes before the patient has begun to get up and move about.

Aortic stenosis, like aortic incompetence, leads up to a fatal termination in two ways—directly by limiting the supply of arterial blood, and indirectly by giving rise to back pressure in the pulmonic and venous circulation, through the intervention of dilatation of the left ventricle. It is when symptoms arise from the latter that digitalis is useful. This drug is even less competent to overcome the direct effects of obstruction than of regurgitation, and the left ventricle may be injured if stimulated to drive its contents through a narrowed orifice. More relief is often obtained by relaxing the arterioles by means of nitro-glycerine, deducting thus the arterio-capillary resistance from the total work with which the heart has to contend.

Mode of Action of Cardiac Tonics in Mitral Regurgitation.

Mitral regurgitation, being the disease in which the action of the cardiac tonics is almost always beneficial, a study of the conditions presented may enable us to arrive at some comprehension of the way in which the good effects are brought about.

What takes place in mitral regurgitation is as follows: The regurgitation into the left auricle dilates this cavity (there may be some hypertrophy of its muscular walls, but no compensatory influence of any consequence is gained thereby) and at the same time drives back the blood which is flowing to the left auricle and ventricle by the pulmonary veins. The obstruction in the pulmonary veins necessarily gives rise to resistance to the onward flow through the capillaries, to overcome which increased pressure is required in the pulmonary artery, and therefore greater driving power on the part of the right ventricle. From this results hypertrophy of the right ventricle, which is the great compensating agency by which the leakage of the mitral valve is more or less perfectly neutralized. If the blood pressure in the pulmonary veins could be maintained at such a point as to be greater than the pressure in the aorta, there would be no reflux into the auricle during the contraction of the ventricle, even were the mitral valve completely destroyed; but, however powerful the action of the right ventricle, this can never be absolutely the case. The walls of the pulmonary capillaries and veins, and of the left auricle are too weak to resist such a distending force, and moreover the suction action of the left ventricle during diastole will always temporarily reduce the pressure in the left auricle.

Another change in the heart resulting from mitral regurgitation must be noticed. This is a dilatation of the left ventricle, produced by distension of this cavity during the defenceless diastolic period, by the high pressure in the pulmonary veins and left auricle. It involves some consecutive hypertrophy of the ventricular walls.

These familiar and elementary explanations are enumerated in order once more to emphasize the fact that the work of compensation for mitral regurgitation falls upon the right ventricle, and that, when systemic venous stasis and other late effects of mitral regurgitation show themselves, it is because the right ventricle is beaten by the resistance in the pulmonary circuit and can no longer keep up adequate pressure in the left auricle.

Applying now our knowledge of the physiological effects of digitalis, we shall see that the favourable results of its administration are due almost entirely to reinforcement of the right ventricle. On the left side of the heart and in the systemic circulation there will be produced a certain degree of arterio-capillary contraction, with slight increase in the peripheral resistance and in the intermediate arterial tension, and a more deliberate and energetic action of the left ventricle in systole, which makes room for a large volume of blood in diastole, while the elastic rebound at the end of systole exercises a better suction action on the contents of the distended left auricle. The hypertrophy of the ventricular walls, which will more than neutralize the increase of resistance in the peripheral circulation, and the greater capacity of the cavity would, in the absence of regurgitation into the auricle, result in the projection of a larger charge into the arteries at each systole. The effect of this is undoubtedly good, but the regurgitation into the left auricle is a set-off against it, and this will be increased with the increase of resistance in the arterial system. So far, however, as these good effects on the left ventricle and systemic circulation are concerned, they would be much more conspicuous in aortic regurgitation than in mitral regurgitation, since the ventricle is stronger and its capacity larger, and yet we do not find that digitalis is more useful in this affection, but very often the contrary.

Looking now at the effects upon the right side of the heart and the pulmonary circuit, there may or may not be contraction of the arterioles and capillaries in the lungs with increase of resistance. This could, however, in any case only be slight, while the ventricular walls being greatly hypertrophied, increase of vigour in their contraction will at once raise the blood pressure in the entire pulmonary circulation and in the left auricle. Improved pressure in the left auricle, as has been seen, will fill the left ventricle better during diastole, will resist reflux through the mitral orifice in the systole, and so will increase the amount of blood thrown into the aorta.

It is here that the beneficial influence of digitalis really comes in. As has been before stated, the neutralization of the effects of mitral regurgitation is almost entirely the work of the right ventricle, and it is by increasing the efficiency of its compensatory action that digitalis is of service. Additional evidence of this is, on the one hand, the fact that relief of over-distension of the right side of the heart by venesection or leeches and purgation is an important corroborative measure, in many cases absolutely essential to the result, and, on the other hand, that digitalis fails when the right ventricle is seriously degenerated or hampered by pericardial adhesions.

Among the conspicuous favourable results of the administration of digitalis is diminished irregularity of the pulse. This is entirely due to the higher blood pressure in the left auricle, and the more regular supply of blood to the ventricle. Mitral incompetence is the one among the valvular affections which is specially liable to give rise to irregularity of the pulse. This will be understood from the following considerations. The left auricle is exposed to the respiratory variations of pressure which its thin walls resist only imperfectly. When, therefore, these variations are exaggerated, as when the breath is held or when there is dyspnœa from bronchitis or asthma, the amount of blood carried on into the left ventricle will vary, and the pulse will be more or less irregular. During inspiration the negative pressure will tend to keep the auricle dilated, and to prevent it from contracting properly, so that the ventricle will not have a full charge of blood, and its systole will be brief and abortive. During expiration the auricle will be compressed, and its contained blood will be forced on into the ventricle, with opposite consequences. In mitral regurgitation a further effect will be that the negative pressure of inspiration will encourage the reflux into the auricle, while the positive pressure of expiration will oppose it. In this way—the ventricle sometimes being imperfectly, at other times perfectly, filled, sometimes sending back more, sometimes less, of its contents into the auricle—we have ample explanation of irregularity of the pulse. Now, the lower the internal blood-pressure in the auricle and pulmonary veins, the greater will be the effect of variations of external pressure, and the higher the pressure within the auricle, the more independent it will be of pressure from without. It will be seen, therefore, how digitalis steadies the action of the heart and renders the pulse more regular.

Effects of Digitalis in Mitral Stenosis.

If we examine the effects of digitalis in mitral stenosis, we may perhaps see why they are less certainly favourable, and sometimes clearly unfavourable. In an uncomplicated case the left ventricle is neither dilated nor hypertrophied, and the arteries generally are already small and contracted. No obvious advantage can be seen in further contraction of the arterioles, and, in point of fact, the symptoms are somewhat relieved by causing them to dilate. No great improvement in the output of blood, again, is to be gained by more vigorous contraction of the walls of the left ventricle, as they are not specially strong, and the cavity is small. But it would seem that increased vigour in the contraction of the right ventricle should have the same good effect here as in mitral regurgitation, and some beneficial influence is indeed very commonly observed at first. The conditions, however, are different. In mitral regurgitation the increased amount of blood driven by the right ventricle into the pulmonary artery by raising the pressure in the pulmonary circulation antagonises the reflux into the auricle, so that more blood finds its way into the left ventricle, whereas in mitral stenosis the blood cannot be forced through the constricted mitral orifice beyond a certain rate of speed, and if the right ventricle is stimulated to contract more than is required for this, it encounters an insuperable obstruction, and becomes embarrassed in its action, its energy being uselessly expended. A common result is irregularity in the beats, accompanied by a sense of precordial oppression, and not infrequently the heart-beats are in couples, the first of which alone reaches the wrist, the second having no aortic second sound.

In many cases of advanced mitral stenosis, the coupled beats can be produced at will by giving digitalis. The second of the two beats is evidently a supplementary systole of the right ventricle: there is a right ventricle impulse felt over the lower left costal cartilages, while the apex beat is scarcely, or not at all, perceptible. At the second of the coupled beats both right ventricle sounds are heard, while the aortic second sound is absent, and if there is a systolic mitral murmur as well as the presystolic, it is audible only with the first of the two beats.

CHAPTER X.

The Individual Valvular Lesions.

AORTIC STENOSIS.

THE MURMUR OF AORTIC STENOSIS — CONDITIONS OTHER THAN AORTIC STENOSIS WHICH MAY GIVE RISE TO SYSTOLIC AORTIC MURMURS—CAUSES OF AORTIC OBSTRUCTION—DIFFERENTIAL DIAGNOSIS OF MURMURS —ESTIMATION OF EXTENT OF LESION BY MEANS OF THE MURMUR, THE CHANGES IN THE HEART, THE PULSE— PROGRESS OF THE DISEASE : SYMPTOMS — PROGNOSIS— TREATMENT.

Aortic Stenosis.

The first valvular change to be considered will be aortic stenosis. This is the least common of the valvular affections of the left side of the heart, as shown both by post-mortem statistics and clinical experience.

The characteristic murmur to which aortic stenosis gives rise is a murmur systolic in time which usually has its maximum intensity in the second right intercostal space close to the sternum, in the so-called aortic area. All systolic murmurs heard over this area do not, however, necessarily imply actual obstruction of the aortic orifice, but may have other significance.

Description of the Murmur.—The systolic aortic murmur, as is indicated by its name, is heard during the systole of the left ventricle, and is produced by the rush of blood through the obstructed aortic orifice from the ventricle into the aorta. Its commencement coincides in time with the first sound of the heart, and may either accompany or replace it, and it may begin with a burst or accent, or gradually. The duration of the murmur varies; it is usually long, sometimes occupying the entire interval between the first and second sounds, which interval may be prolonged, but it may be short. It is audible over the sternum at the level of the third intercostal space, but is most distinct just outside the right edge of the sternum in the second space, or sometimes over the third costal cartilage, at which point on the surface the aorta comes from under the pulmonary artery, and nearly touches the anterior wall of the chest. It is conducted upwards, as the phrase is, along the right side of the sternum to the right sterno-clavicular articulation, where it is distinctly audible, and it is also heard over the carotids in the neck and occasionally over the thoracic aorta along the spine. It is frequently heard along the right margin of the sternum lower down, sometimes to the fourth or fifth space right or left of the sternum, more especially when the aorta is dilated and elongated, and it may have its maximum intensity over the sternum near its left edge at the level of the third rib or space, i.e. immediately over the aortic valves. This, however, is not common, as the root of the aorta is deeply seated in the chest, and has, between it and the surface, the conus arteriosus of the pulmonary artery. The murmur is lost over the right ventricle, but is again audible, as a rule, at the apex, being conducted to this point by the wall of the left ventricle. This murmur may be loud—sometimes extremely loud—and when this is the case, it may be heard all over the chest behind, as well as in front, or it may be comparatively soft; it is often rough and vibratory, more rarely, croaking in character, but it may be smooth and blowing, or musical. Occasionally it is accompanied by a systolic thrill felt in the second or third right space, or in both close to the edge of the sternum.

Causes other than Aortic Stenosis which may give rise to a Basic Systolic Murmur.

Such a murmur in one or other of its varieties may be due to several causes, which will now be enumerated.

1. Anæmia may be mentioned first. It is not well understood how it is that watery blood, deficient in corpuscles, gives rise to murmurs, but the fact is exemplified by the venous hum heard in the neck, by the pulmonic murmur, and also, though less commonly, by a systolic aortic murmur. This murmur is rarely rough or loud, but it is often distinct and audible along the edge of the sternum upwards to the sterno-clavicular articulation, and in the neck. It accompanies the first sound, and does not substitute itself for it.

With this anæmic aortic murmur may be grouped the hæmic murmur of very similar character, sometimes accompanying or following acute febrile disease of long duration such as rheumatic fever and typhoid fever, although it is not certain that the causation is exactly the same. It is not very uncommon, or rather it was not before the employment of salicin and the salicylates to hear, after an attack of acute rheumatism, a systolic murmur at both mitral, pulmonic, and aortic orifices, which would entirely disappear as strength was regained, and the presence of multiple murmurs is a reason for attributing them to the state of the blood, and to the innutrition of the tissues left behind by the acute disease.

2. Mere roughening of the orifice or valves, or impaired flexibility of the latter, slight congenital malformation, fenestration of one or more of the cusps, a shred of fibrin

hanging from the edge of a valve, may give rise to a loud systolic murmur without offering obstruction to the course of the blood.

3. An aortic murmur may be produced by acute or sub-acute aortitis, a rare and obscure disease which may be suspected when, with a murmur not previously known to be present, there are irregular pyrexia, dull sub-sternal pain, and rapid and apparently unaccountable failure of the heart.

4. Dilatation of the aorta just above the valves, the orifice and valves remaining unchanged, produces a murmur. The fibrous ring at the root of the aorta, which gives attachment to the muscular fibres of the ventricle, and supports the semilunar valves, is extremely strong, so that it does not readily give way and allow the orifice to be enlarged, but the aorta just above the ring is prone to dilatation, and the blood passing through an opening of normal size into a larger cavity beyond is thrown into eddies, the vibrations attending which produce the sound heard as a murmur.

In none of these is there stenosis of the orifice, or obstruction to the stream as it issues from the heart; the condition last mentioned has dangers of its own, but they do not arise from aortic obstruction.

Stenosis of the Aortic Orifice.—Actual narrowing of the mouth of the aorta is almost always due to changes in the valves, the cusps of which may be stiff and rigid, with rounded and thickened margins, a condition which does not permit them to fall back into the sinuses of Valsalva before the current of blood. Instead, therefore, of a circular orifice of the same size as the vessel beyond, there is a roughly triangular opening of reduced area, formed by the edges of the cusps: vegetations so-called may further encroach on this space. Sometimes adhesions take place between the valves at the angle in which they meet, which further narrow the orifice. This condition may be left by endocarditis, whether rheumatic, or occurring in the course of scarlet fever or other acute disease. Or, without being greatly thickened, the valves may be affected by atheroma or chronic degenerative changes, which destroy their flexibility, and cause puckering and contraction of their margins, or lead to calcareous deposits in their substance.

It is also possible that by dilatation of the ring at the root of the aorta, a slight and unimportant obstruction may be produced; valves, not themselves greatly diseased, being so put on the stretch that they cannot fall back, but remain as an obstacle to the stream of blood.

Differential Diagnosis of Basic Systolic Murmurs.

It is clear that the first point to be ascertained when a systolic murmur is present is, whether it indicates the existence of obstruction or not. In this we shall be greatly assisted by the history, aspect, and age of the patient. We should, for example, suspect that the murmur was hæmic, and not produced by obstruction if the patient were a young and anæmic girl, or if it were first heard during convalescence from an acute illness, more especially if similar blowing murmurs were heard at other valves as well.

The history of an attack of rheumatic fever would on the other hand favour the conclusion that the orifice was actually narrowed.

A systolic murmur over the aorta, appearing after middle age, or in advanced life, or discovered for the first time at this period, will seldom be due to actual narrowing of the orifice, but will be caused by roughness or rigidity of the valves, or by dilatation of the aorta above the valves with perhaps atheromatous irregularities in its walls. Fortunately the aortic second sound is of very great assistance in determining the point. Under the condition just mentioned it will almost certainly be unduly loud and accentuated from co-existing high arterial tension, while stenosis diminishes the intensity of the second sound in two ways: by the changes present in the valves impairing their flexibility, and by the less sudden recoil which follows the slower distension of the arterial system. Prolonged observation of cases of high arterial tension from gout, renal disease, or other causes in advanced life will not unfrequently afford the medical attendant the opportunity of noting the first appearance and gradual development of a murmur over the aorta in addition to the accentuated second sound.

The loudness or character of the murmur, or the extent to which it is conducted along the aorta, does not give us much help. Hæmic aortic murmurs are usually soft and smooth, but so also may murmurs be which are due to extreme narrowing. It has been stated by some that when the margins of the thickened valves are smooth and rounded there may be no murmur in spite of considerable narrowing; but if the fact be true, the explanation can scarcely be accepted, since it is not the friction of the blood against the valves as it passes over them which generates sonorous vibrations, but the eddies produced by the passage of the blood through a small orifice into a large channel. A more probable explanation is weakness of the ventricle and languid propulsion of the blood which may render the murmur weak or inaudible; thus the gradual enfeeblement and ultimate disappearance of a systolic aortic murmur may be a most serious prognostic indication. The very loudest murmurs, such as have already been mentioned, are most frequently produced by a fringe hanging to the edge of one of the cusps, or by some roughening or calcareous patch, or by a delicate band stretched across a part of the orifice, but it would not be safe to conclude that all loud murmurs are harmless. All we can safely infer is that a long, loud murmur indicates a vigorous ventricular systole, which is a good prognostic element.

It is thus evident that from the murmurs alone it may be impossible to make a certain diagnosis of aortic stenosis, and still less to estimate the amount of the constriction.

Estimation of Degree of Stenosis by Degree of Cardiac Hypertrophy.

We turn, then, to the condition of the heart. If there is actual obstruction its existence will be betrayed either by the changes in the walls and cavities which are required in order to overcome it, or by some evidences of derangement in the circulation.

The change by which aortic obstruction will be overcome will be more or less pure hypertrophy of the left ventricle. This does not bring the heart to the gigantic size which it reaches as a consequence of aortic regurgitation. The apex beat will be lower than the normal position by an intercostal space, or perhaps more, but it will not be greatly displaced outwards; it will be a well-defined and deliberate push of no great violence. The first sound here will be dull and prolonged, and not very loud; perhaps accompanied by a murmur which may be mitral, but most commonly is, in the early period of the disease, the aortic murmur conducted to the apex by the walls of the left ventricle. The other sounds will present nothing remarkable, except that the aortic second sound will be weak. If we get such evidence of hypertrophy of the left ventricle as this, *i.e.* the downward displacement and definite push of the apex with a prolonged first sound, in a young person without kidney disease, together with a systolic aortic murmur, there can be no hesitation in inferring stenosis of the aortic orifice. At and after middle age it would be necessary to consider whether the hypertrophy might not have been produced by long-continued, antecedent resistance in the peripheral circulation and high arterial tension; but in this case the apex would usually be less defined, the first sound louder and less deliberate, and the aortic second sound much accentuated.

The Pulse.

Almost more important than the hypertrophy of the left ventricle will be the character of the pulse. The orifice being narrowed, the blood discharged from the ventricle will require more time to pass through it, and the pressure in the arterial system will not at once rise to its maximum; in other words, the so-called percussion element of the sphygmographic trace will be weakened and the pulse wave will be long and slow, not striking the finger or lever abruptly and vigorously, but raising it gradually. The artery will, for the most part, be small and full between the beats of the pulse. A large sudden pulse is incompatible with aortic stenosis unaccompanied by regurgitation, and when it is a question whether or not an aortic systolic murmur is due to obstruction at the orifice, the pulse becomes the most certain criterion to which we can appeal; if there is an apparent contradiction between the indications of the heart and those of the pulse, the latter must dominate.

For instance, if with a systolic and diastolic aortic murmur and a hypertrophied heart, the pulse is large, sudden and collapsing, we shall infer that there is no real stenosis of the aortic orifice; if, on the other hand, the pulse lacks these features characteristic of aortic incompetence, we shall conclude that there is aortic stenosis which has interfered with their development; and we shall estimate the degree of stenosis from the extent to which it has modified the pulse.

A systolic murmur heard over the aortic valves and along the aorta will not, in the absence of cardiac hypertrophy and the long pulse described above, indicate stenosis of the aortic orifice, and even when there is hypertrophy, if the pulse is large, short and abrupt, there can be no real narrowing.

Progress of the Disease. Mode of Death.

When the constriction is moderate in amount, and is a result of endocarditis, in a young subject the heart will usually undergo hypertrophy sufficiently to overcome the obstruction, and no serious or troublesome symptoms will arise. There is a tendency, however, for the constriction to increase, and when this is so considerable as to have given rise to great hypertrophy of the left ventricle, dilatation is sure to follow sooner or later. As a consequence of the deficient propulsive power resulting therefrom, there will be a tendency to systemic stagnation, and as a result of the incomplete emptying of the ventricular cavity in systole, there will be insufficient room for the whole contents of the left auricle in diastole, and consequent back pressure through the left auricle and pulmonary circulation. Mitral incompetence may ensue as a result of dilatation of the left ventricle, but there will also be other injurious influences at work which may give rise to it by damage to the mitral valve in the following way: The pressure in the ventricular cavity will be greatly increased owing to the powerful contraction of the muscular walls, which is necessary to overcome the obstruction at the orifice; as a consequence of this, there is a constant and severe strain upon the flaps of the mitral valve and the chordæ tendineæ, so that eventually they become stretched or give way, and mitral incompetence results. It will thus be seen that in aortic stenosis of any severity there is little chance of the mitral valve escaping damage in the long run, and if it has at the outset been injured by the same attack of endocarditis which gave rise to the aortic mischief, the outlook is far more serious. The barrier formed by the mitral valve being thus broken

down, the door is open to backward pressure through the
auricle, which will take effect upon the pulmonary circulation and right ventricle. It has been said by some that
timely yielding of the mitral valve acts as a safety valve,
as tricuspid regurgitation is supposed to do for the right
ventricle, preventing sudden death through over-distension
and consequent paralysis of the left ventricle. This view
does not commend itself to my judgment. It must, in the
nature of things, be purely hypothetical, and no basis for it
is found in the occurrence of sudden death produced in the
way described. However this may be—and in my opinion
the hypothesis has no foundation in fact—the establishment of mitral regurgitation marks a downward step and
renders the prognosis grave. The obstacle to the outflow of
blood from the ventricle, which has given rise to the whole
train of consequences, remains irremovable, and the setting
in of mitral symptoms marks the failure of compensation.

PROGNOSIS.

Nearly all authors concur in making aortic stenosis the
least dangerous of the valvular affections, and on *a priori*
considerations such would seem to be probable; but this
conclusion will scarcely seem quite secure if the cases of
systolic aortic murmur without actual narrowing of the
orifice are eliminated. Just as the inclusion of all cases in
which a systolic aortic murmur is heard makes aortic stenosis
apparently the most frequent valvular disease, while on
post-mortem evidence it is seen to be the least common, so
the dilution of the death-rate by these cases in which no
real narrowing exists makes it appear to be the least fatal.
The relative danger of real stenosis cannot be estimated
with confidence, but it is certainly greater than has been
supposed, and though it is not so serious as aortic incompetence or mitral stenosis, it is more so than mitral incompetence. The average age of the cases of this disease

in the post-mortem statistics previously referred to, was about forty, and would have been higher had deaths by intercurrent independent disease been excluded. The age of the oldest patient among these, however, was fifty-three; this would bear out the above statement that aortic stenosis must be looked upon as more dangerous than mitral incompetence, of which numerous examples are known to have reached the age of seventy.

There is no risk of sudden death as in aortic incompetence, and there is little danger as long as the patient is free from symptoms; but when once dropsy has set in from back pressure due to dilatation of the left ventricle, there is a smaller possibility of recovery than in other valvular affections. For when the left ventricle has given way under the resistance it encounters, it has little chance of regaining its normal condition, since the resistance persists undiminished, and there is no other compensatory influence which can be called upon. Further, if the break-down occurs after middle life, it is probable that degenerative change is taking place in the hypertrophied left ventricle, which will render the prognosis still more unfavourable.

In regard to this it must be borne in mind that at, and after middle age, a systolic aortic murmur appearing for the first time will be indicative of degenerative change, and obstruction is the very smallest of the dangers to which degeneration in or about the root of the aorta may give rise. A damaged valve may yield and permit of regurgitation, or the degeneration may affect one of the sinuses of Valsalva and produce aneurysm here, or may be part of a more general atheromatous change extending over the entire ascending aorta, or, what is more serious, it may implicate the orifice of the coronary arteries, and by cutting off the supply of blood lead to fatty change in the walls of the heart. Any murmur at the aortic orifice, then, in an elderly person, must be looked upon as a possible forerunner of serious disease.

TREATMENT.

In regard to the treatment of aortic stenosis, there is little to add to what has been said on the treatment of valvular disease in general. We cannot hope to rectify the constriction which has already taken place, and can do little to modify the process of narrowing, which may be going on as a result of cicatricial contraction, or adhesion of the valvular curtains, after acute endocarditis has subsided. When, however, the valves are being damaged by a chronic inflammatory process due to syphilis, striking benefit may be sometimes obtained by the administration of large doses of iodide of potassium.

Supposing the contraction of the aortic orifice to have reached a certain point and become stationary ; the dangers likely to arise out of this condition will be, at an early period, an imperfect degree of compensatory hypertrophy and later degeneration of the hypertrophied muscular wall, with, in both cases, secondary dilatation of the left ventricle. From this, mitral incompetence may result, and subsequently right ventricle failure with its train of symptoms of venous obstruction.

The hypertrophy is usually sufficient in the first instance, as the contraction of the orifice takes place slowly, so that the structural increase can keep pace with it, provided that the patient is a young subject. A moderate degree of constriction may therefore exist for years without apparently affecting the health or well-being, or interfering with ordinary occupations, giving rise to no symptoms except perhaps shortness of breath and præcordial pain on too great exertion. Even considerable stenosis is not incompatible with moderate exercise and work and apparent health.

Exceptions occur when the patient is weak and anæmic, either as a result of protracted illness, or from other causes, and nutrition being poor, the hypertrophy is inadequate and accompanied by dilatation. In such cases prolonged rest may be of great service.

Secondary dilatation is to be averted by carefully avoiding over-exertion, fatigue, and anxiety, and by attention to general health. Moderation in food and drink is necessary, as the amount of exercise which can be taken is limited, and anything that may lead to high arterial tension and increased peripheral resistance, is especially to be guarded against. Tonics may be given when required, and anæmia is to be combated by all means in our power. The protracted administration of digitalis is of very questionable utility. When the left ventricle has begun to give way, and symptoms of backward pressure through the lungs and embarrassment of the right ventricle supervene, then digitalis and similar remedies will be of service. If, however, these symptoms have come on insidiously, and are not traceable to over-exertion, chill, or any definite cause, drugs will rarely be able to improve materially the condition of the patient or postpone for long the fatal termination.

When the hypertrophied walls of the heart have begun to undergo degeneration, and there is præcordial pain and oppression, the administration of nitro-glycerine and nitrites, or the nitrates of erythrol and mannitol, by diminishing the arterio-capillary resistance, and thus relieving the stress on the left ventricle, will often afford the heart distinct relief, and produce a general amelioration of symptoms. These drugs may be given at the same time with digitalis.

On the other hand, I have more than once known aconite, given with a similar object, completely overthrow the compensation established by hypertrophy, and cause rapid cardiac dilatation, with frequent weak and small pulse accompanied by pallor and cold sweats.

CHAPTER XI.

AORTIC INCOMPETENCE.

PHYSICAL SIGNS—THE DIASTOLIC MURMUR: DIRECTIONS IN WHICH IT IS CONDUCTED—PRESYSTOLIC MURMUR, ITS SIGNIFICANCE—MODIFICATION OF THE AORTIC SECOND SOUND—PULSATION OF ARTERIES—CAPILLARY PULSATION—THE COLLAPSING PULSE—PULSUS BISFERIENS—IRREGULAR PULSE—ESTIMATION OF THE AMOUNT OF REGURGITATION BY THE CHARACTER OF THE MURMUR, OF THE AORTIC SECOND SOUND, OF THE PULSE, AND BY THE CHANGES IN THE HEART—AORTIC INCOMPETENCE DUE TO SYPHILIS AND CAUSES OTHER THAN ACUTE ENDOCARDITIS—SYMPTOMS—PROGNOSIS—TREATMENT.

THE diagnostic sign of this condition is a murmur produced by the backward rush of the blood from the aorta into the left ventricle during its diastole. Two forces will be at work in its production—the suction action of the ventricle, and the pressure in the arterial system generally and in the aorta in particular, due to the elastic recoil of the walls of the great vessels. The cause of the murmur is not necessarily roughness or irregularity of the valves or orifice; it may be due partially or entirely to the vibrations generated by the passage of liquid through a constricted point in a channel, into a large cavity beyond.

Physical Signs of Aortic Regurgitation.

1. **Murmur.**—The seat of production of the murmur being the valves which encircle the root of the aorta, the point on the surface of the chest immediately over and nearest to the origin of the sound will be close to the left margin of the sternum at the level of the third cartilage; but it is not always here that the murmur is best heard, as the conus arteriosus of the pulmonary artery is interposed between the aorta and the chest wall.

The murmur is conducted in various directions, and the seat of its maximum intensity differs in different cases. It is, with rare exceptions, audible in the so-called aortic area just outside the right edge of the sternum in the second intercostal space, or over the second costal cartilage, and is usually conducted upwards from this point along the sternal margin as far as the sterno-clavicular articulation, though losing rapidly in intensity the while. It can frequently be followed downwards along the side of the sternum to the fourth space, or even lower, sometimes with increasing intensity over one or two ribs; it is also frequently heard along the sternum itself, or to the left of it, the seat of maximum intensity being sometimes in the fourth or fifth spaces near the left edge of the sternum. It may again be audible at the apex after having been lost over the right ventricle, but it is usually weakened here, and is never at its loudest. Another point at which it may be heard more distinctly than anywhere else, is the third left space exactly over the pulmonic valves or artery; it may indeed be limited to a small area in this region.

Cause of the Conduction of the Murmur.—In a certain degree we may say that the distribution of the murmur described is explained by its conduction downwards by the current of the regurgitant blood, and this must be held to account for the difference between obstructive and regurgitant aortic murmurs, in regard to the points on the chest wall where they are best heard. But the direction which the blood takes will be towards the apex, and if the current carried the murmur, it would be at the apex that we should hear it most distinctly, which is not the case. The truth is that we are apt to lose sight of the fact that much more blood enters the ventricle from the auricle than from the aorta—any other state of things would be incompatible with onward movement of the blood in the systemic arteries, and therefore with life—and also that the regurgitant stream through the valves is comparatively small if rapid. We are also perhaps given to figure in our imagination, this stream as falling into an empty vessel, whence has arisen such an explanation as the collision of the blood with the ventricular wall, to account for some abnormal sound or other. The ventricle is never an empty cavity into which blood can fall with violence as into a bottle or jar; there is no large and powerful stream rushing straight from the aorta down to the apex; during diastole in a case of aortic insufficiency, there must be complicated cross currents in the ventricle, rendering conduction of sound in any definite direction impossible. The vibrations, then, must be conveyed by the walls of the ventricles, and not by the contained blood, or at any rate the blood thrown into sonorous eddies will communicate the vibrations to all parts of the ventricular wall indifferently, and not impinge upon any special point carrying thither the sound.

Presystolic Murmur.—Sometimes, more usually when there is stenosis as well as incompetence of the aortic orifice, a presystolic murmur is heard in the region of the apex, resembling very closely that of mitral stenosis, which does not, however, denote constriction of the mitral orifice. It is, of course, possible that mitral stenosis may co-exist with double aortic disease ; but in numerous cases in which a presystolic murmur has been audible at the apex, together with a systolic and diastolic murmur in the aortic area, the mitral orifice has been found to be normal at the autopsy, and the only lesion has been constriction and incompetence of the aortic orifice. The following explanation of the origin of this murmur may be suggested. The mitral and aortic orifices are in close apposition, and it is probable that the regurgitant stream from the aorta impinges to a certain extent on the anterior or aortic flap of the mitral valve. It is conceivable that the murmur may be due to this flap of the mitral valve being thrown into vibration by the regurgitant stream from the aorta, the vibration being conducted to the apex by the chordæ tendineæ and papillary muscles, or, on the other hand, that the backward rush of blood from the aorta prevents the complete falling back of this flap of the mitral valve, so that some actual obstruction of the mitral orifice results, giving rise to a presystolic murmur.

Violent Arterial Pulsation.—One of the most striking features in aortic regurgitation is the violent pulsation of the arteries throughout the body, most conspicuous in the carotids. So marked is this that one is often compelled to make an involuntary diagnosis in church or on a casual meeting in the streets. The pulse is visible at the wrist, the brachials at the bend of the elbow throw themselves into violent curves, the femoral and anterior tibial arteries are scarcely less visible; the temporals, facials, and labials may be seen beating; even the digital arteries may be seen as well as felt to an unpleasant degree. As was first pointed out by Sibson, the radial pulse is audible when the hand is raised, a sign more interesting perhaps than valuable.

The arteries, as may be easily seen in the radial at the wrist, are large, because the entire contents of the left ventricle, expanded to two or three times its normal capacity, are launched into the arterial system, distending it momentarily to a corresponding degree.

Capillary and Venous Pulsation.—Another interesting effect is capillary pulsation, a pulsatile reddening of the skin, which is sometimes observed in the palms of the hand when warm, especially when pyrexia is present, and a similar phenomenon may at any time be provoked at various parts of the surface, most conveniently perhaps on the forehead, by bringing out a red patch or line by friction, the margins of which will be seen alternately to extend and fade, synchronously with the pulse. This is due to the extension of the arterial relaxation into the capillaries, and when it is general the pulsatile movement of the blood may even reach the veins. To render this visible the hand should be so held as to drop at the wrist, when the veins on the dorsum will fill, and sometimes will be seen to pulsate, not with the sudden beat of the artery, or like the rapid flush of the capillaries, but with a gentle and deliberate movement followed with difficulty, but which a filament of sealing-wax across the vein will render visible.

Capillary pulsation, however, is not a phenomenon peculiar to aortic incompetence, as it may be produced sometimes in phthisis or enteric fever, or other conditions associated with a pulse of low tension.

The Pulse.—This is the well-known collapsing or water-hammer pulse, sometimes named after Corrigan, who is believed to have first called attention to it. The artery at the wrist is large. In the intervals between the pulsations it is empty and allows itself to be completely flattened against the bone; then the wave comes with a sudden violent rush, filling the vessel and lifting the fingers forcibly. It is as short as it is sudden, and the artery at once collapses again under the pressure of the fingers. In order that all these features may be fully realized, the hand must be above the level of the heart and above the shoulder or elbow. When the patient is lying in bed, the act of giving the hand for the pulse to be felt brings about this condition, but when he is sitting, as during an interview in the consulting-room, or standing, although the artery may be large and the beat sudden, the collapse will not take place.

The cause of the collapse is, that in consequence of the loss of the support of the aortic valves, the column of blood is not sustained, and therefore drops out of any vessel which is above the level of the heart in obedience to gravity. If, then, the hand is hanging down, the radial artery remains full, having above it a column of blood up to the arch of the sub-clavian.

The collapsing character of the pulse is indispensable as an evidence of aortic regurgitation, though certain complications, which will be enumerated later, may interfere with the degree of collapse.

In cases of pulmonic regurgitation, which are rarely met with, the absence of the collapsing pulse and of undue carotid pulsation will be the most important distinguishing features. Similarly their presence, in cases where the diastolic murmur is heard loudest or exclusively over the pulmonic area, will clear up all doubts as to diagnosis.

The Delay of Pulse.—The pulse in aortic regurgitation is always retarded or delayed—that is, there is an appreciable interval between the beat of the heart, or the carotid pulse, and the pulsation in the radial artery, which varies according to the extent of the incompetence. This delay is due partly to the collapsed and empty state of the arteries between the beats, and partly to their large size and loss of tone. There is no longer a continuous column of blood between the heart and periphery, and the tension in the arterial system has to be considerably augmented by the launching of a large volume of blood into the aorta, before the vessels are rendered sufficiently tense for a pulse-wave to be transmitted.

The **pulsus bisferiens** is sometimes met with in aortic incompetence, more commonly when there is concomitant stenosis. It is a peculiar double beat, best felt when the fingers exert a moderate pressure on the artery, less than is necessary to bring out fully the collapsing character of the pulse, but more than is employed to appreciate dichrotism. It can be readily demonstrated by the sphygmograph. It is produced by a double systolic effort, which can sometimes be felt or heard in the heart itself, and which frequently gives rise to a double rush of blood audible in the carotids. According to D'Espine, of Geneva, the normal systole of the heart is à deux temps, or a double contraction, of which this is an exaggeration. The pulsus bisferiens can not uncommonly be induced by an effort, which throws additional work on the heart; for instance, in one case it was not present while the patient lay quietly in bed, but was brought out when he held up both his hands.

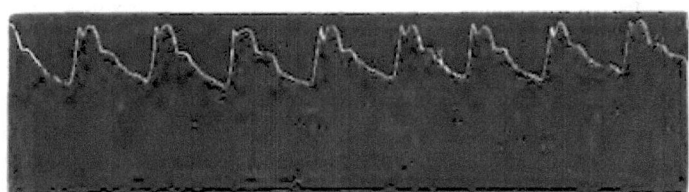

FIG. 8.—PULSUS BISFERIENS.

Irregularity of Pulse.—Though the pulse of aortic incompetence is for the most part regular in force and frequency, in advanced cases, especially when the heart is beginning to fail, irregularity of pulse is not uncommon. The irregularity is first manifested by the occurrence of a short and rapid pulsation of less force and amplitude than the ordinary wave, and occurring at irregular intervals. It succeeds the previous beat very quickly, and is followed by a longer interval than usual. It would seem to indicate an effort of the heart to supplement an inadequate and inefficient preceding contraction, or to be an abortive systole. Later on, the irregularity, both in force and frequency, becomes more general and more marked, though the pulse is still recognizable as that of aortic regurgitation.

Estimation of the Extent of the Lesion.

The amount of blood which returns into the ventricle is a matter of the first importance in estimating the prospects of life and comfort before the patient, and we have now to consider the different sources of information on this point.

The diastolic murmur itself affords but vague indications. As has been already mentioned, it may be loud or feeble, rough or smooth, long or short. It usually begins with an accent. Speaking generally, a long loud murmur shows that a considerable degree of pressure is kept up in the aorta, which is a desirable thing in itself, and a proof that the heart is acting with vigour, and also that the leakage of the valves is not excessive; it is usually, therefore, among the favourable auguries. On the other hand, a weak short murmur indicating an opposite state of things may be a note of impending danger or death. But there are so many exceptions to the rule hinted at that it is not to be relied upon.

Aortic Second Sound.

A very important auscultatory sign, however, is the presence or absence of the aortic second sound. We must listen for this, not at the apex or in the aortic area, but in the neck. A second sound of some kind, probably the pulmonic conducted, is often heard at the apex or base, but it has not the same favourable significance. The point of the sign is this: the aortic second sound is produced by the sudden tension of the aorta and its semilunar valves at the moment of closure of the valves; it is not their clicking as they meet, or the tension of the valves alone under the column of blood, but the vibration of the entire ascending aorta. If, then, the incompetence is considerable, there cannot be the sudden check to the column of blood which sets the aorta vibrating, and the diastolic murmur takes the place of the second sound; if, on the contrary, the leakage is only small, the required check or shock is given by the closing valves, and the second sound is distinct, although there may be a murmur. It is not the murmur, which may be loud or feeble, that drowns the second sound and prevents it being heard. On listening in the neck over the carotid artery, we have the advantage of being out of reach both of the diastolic murmur and of the pulmonic second sound, and the tension in the aorta must be real in order that the second sound may be heard here. A second sound, therefore, heard in the neck indicates that the regurgitation is small in amount and is consequently a favourable prognostic element.

The Pulse.—The degree of collapse in the pulse is an important factor in the estimation of the amount of the regurgitation. The greater the regurgitation, the more pronounced will be the collapsing character of the pulse. In the absence of any marked collapse in the pulse, a diastolic murmur, whatever its character, would not indicate any serious degree of incompetence. While this may be taken as a general rule, three important exceptions must be mentioned:

1. Concomitant aortic stenosis may interfere with the suddenness and completeness of the collapse, when the amount of regurgitation would have to be estimated by other means than by the pulse.

2. In the last stages of aortic regurgitation, when the heart is failing, there may not be sufficient force in the cardiac systole to produce the collapsing pulse.

3. In aortic disease, acquired in later life from dilatation of the aorta or degenerative changes in the valves, the

pulse has not the typical sudden and collapsing character. This absence of collapse in the pulse is partly due to the rigidity and loss of elasticity in the vessels, partly to the fact that the incompetence is not great, as in such cases life is rarely prolonged if the regurgitation becomes considerable.

The difference between the pulse of aortic incompetence, due to degenerative changes in the valves, and the ordinary collapsing pulse is well brought out by a spyhgmographic tracing.

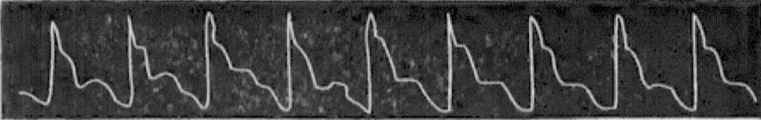

FIG. 9.—COLLAPSING PULSE OF AORTIC REGURGITATION DUE TO ACUTE ENDOCARDITIS.

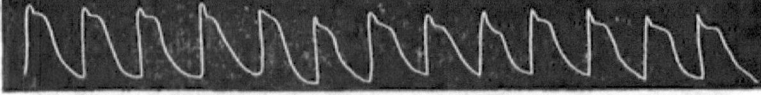

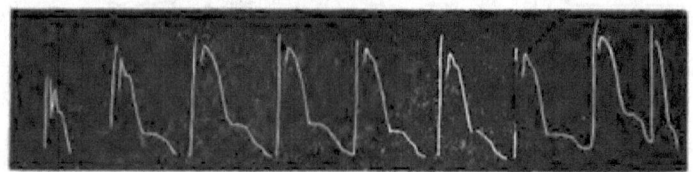

FIG. 10.—LOSS OF COLLAPSE IN PULSE OF AORTIC REGURGITATION DUE TO DEGENERATIVE CHANGES.

The Heart.—Other evidence as to the amount of regurgitation is obtained from the changes which it has induced in the heart. It has already been pointed out in the chapter on dilatation and hypertrophy, that if the normal amount of blood is to be propelled through the systemic arteries, while a certain proportion of that sent by each systole into the aorta is returned into the ventricle, an increase in the capacity of the ventricle is necessary. In other words, it must be dilated. We are not arguing that, because the necessity exists, therefore the provision is made, but showing that the dilatation always found is not injurious but useful, inasmuch as increased force in the ventricular contraction would not alone meet the difficulty.

The process by which the dilatation is effected is the distending influence of the backward pressure from the aorta, during the diastole of the ventricle, when the muscular fibres are relaxed and unresisting. Finally, hypertrophy results from the excessive functional activity imposed upon the walls of the ventricle by the mechanical conditions present. It is in aortic incompetence that the extremes of dilatation and hypertrophy of the left ventricle are met with, and we have the apex beat displaced downwards to the sixth, seventh, or eighth space, and carried beyond the nipple or, in exceptional cases, up to or beyond the anterior axillary line. The apex beat may be well defined, or spread over a large area, and will be powerful. Sometimes the wall of the chest between the apex and the sternum is visibly lifted by the forcible contraction of the left ventricle. Very frequently two or three spaces, between the nipple line and the sternum, will be depressed with the systole, and this systolic recession of intercostal spaces has been supposed to prove the existence of adhesion between the pericardium and heart. It is, however, frequently only the result of atmospheric pressure contributing to fill up the space left by the great diminution of volume of the heart during systole.

ESTIMATION OF THE EXTENT OF THE LESION BY DEGREE OF DILATATION AND HYPERTROPHY OF THE HEART.

The dilatation and hypertrophy being the effects, are also, with certain qualifications, the measure of the regurgitation, in the absence of symptoms such as conspicuous breathlessness on very slight exertion, or faltering action of the heart and tendency to fainting or giddiness. The most important qualification is that arising out of the inability of the heart to take on these compensatory changes late in life; so that when aortic insufficiency is discovered for the first time after the age of thirty, unless its beginning can be distinctly traced and dated, an opinion as to its degree, and as to the danger attending it, can only be arrived at after prolonged observation and repeated examination. It is an interesting and important question up to what age effectual compensation for serious aortic regurgitation is possible. In my own experience all really satisfactory examples have been among those in whom the valvular lesion has been established under the age of twenty. Cases have come under my observation in hospital practice in which the valves have been extensively damaged between the age of twenty and thirty, and the patients have remained under observation for some time afterwards, without developing symptoms of cardiac failure; but such cases are too soon lost sight of for trustworthy conclusions to be based upon them.

The amount of dilatation attending a given degree of incompetence will also vary according to the amount and duration of the care bestowed upon the patient during the first few months of its existence. If he has been allowed to leave his bed too early and engage in exercise, or has undertaken work before the heart has fully recovered from the effects of acute disease, the dilatation will have gone further than would have been the case if due precautions had been observed till the strength was fully restored. Dilatation, moreover, may be increased during subsequent illness attended with pyrexia or anæmia. In all such cases, however, if the extent of the changes in the cavities and walls of the heart does not accurately correspond with the degree of change in the valves, it indicates the extent to which the functional efficiency of the heart is endangered.

Aortic Incompetence due to other Causes than Endocarditis.

In the account just given of the physical signs of aortic incompetence, reference has been made from time to time to differences observed when the valvular disease is established during or after middle life, and from causes other than rheumatism. Such causes are syphilitic affection of the valves and the wall of the aorta, degenerative changes included under the general term atheroma, and rupture of a cusp of the valve.

Syphilis may be suspected as the cause of aortic incompetence, which comes on in early adult life or middle age, and is not traceable to rheumatic endocarditis or to gout or high arterial tension. The regurgitation may rapidly become considerable, and as regards the murmur and the character of the pulse, there will be no important deviations from the description given. The principal difference will arise from the fact that the heart-walls and cavities do not so readily take on compensatory changes; consequently the relation between the degree of dilatation and hypertrophy which makes the latter approximately a criterion by which to estimate the former, no longer holds good. There may, therefore, be considerable regurgitation with comparatively little enlargement of the heart.

When the incompetence is the result of degenerative changes, there will be hypertrophy and dilatation of the heart, not, however, induced by and compensatory of the valvular defect, but developed antecedently to it by the high arterial tension which has given rise to the aortic disease. The pulse will exhibit only partially and imperfectly the collapsing character; the artery will be large, and the wave sudden and brief; but when the hand is held up the vessel does not empty, but can be rolled under the finger throughout the diastolic interval. It will be visible, not however from the sudden filling of the previously collapsed artery, but from the artery being thrown into curves. Further, there will not be the marked loss of time between the apex beat and the pulse at the wrist. The second sound is usually heard in the neck, and has the low pitch and ringing character indicative of dilatation of the aorta and high arterial tension. Very frequently the diastolic murmur is audible across the manubrium, along the course of the aorta.

The insufficiency in such cases is produced, not simply by changes in the valve, but by concurrent dilatation of the aorta implicating its orifice, which is sometimes, indeed, so stretched that the valves, even when retaining their normal size, fail to close it, and the actual amount of regurgitation can never become considerable without producing fatal syncope.

Rupture of a cusp of a valve may occur as the result of severe strain or exertion. The symptoms attending this have already been discussed: their sudden onset and severity are usually diagnostic, and the patient, as a rule, does not long survive the injury.

SYMPTOMS.

The symptoms in aortic incompetence have been fully enumerated in Chapter IV. Those which lead up to a fatal termination fall into two series, according as the ultimate effect on the circulation is arterial stagnation from deficient *vis a tergo* or so-called asystole, or obstruction to the venous return. In the former series, breathlessness, syncopal attacks, anginoid pains, paroxysms of dyspnœa, a sense of oppression in the chest, and præcordial pain are the chief distinguishing features; the pulsus bisferiens is not uncommon, and the special form of irregularity and intermission of the pulse is often present when the symptoms are severe. It is in this class of cases that sudden death is liable to occur from a syncopal attack.

In the other series in which the final symptoms are due to pulmonary and venous obstruction, the last stage does not differ materially from that of mitral disease. The mode of causation of the symptoms is, in fact, almost identical, especially when insufficiency of the mitral valve is an intermediate stage in their production. There must, however, be imperfect filling of the system from the arterial side as well as obstruction in the veins, which will aggravate the tendency to and accelerate the onset of the final and fatal stasis of the circulation.

Mitral regurgitation may be present in association with aortic regurgitation, either as a result of primary inflammation of the valve or as a secondary consequence of the aortic disease which, as has been pointed out, may give rise to it in at least two ways: either by inadequate contraction of the auriculo-ventricular orifice during systole, so that the flaps of the valve, though of normal size, are not large enough to close it; or by thickening and puckering of the valves and tendinous cords, the effect of chronic inflammatory process set up by the increased strain upon them. When once serious inadequacy is produced, its effects will gradually develop themselves in backward pressure upon the pulmonary veins with consequent congestion of the lungs, cough, and shortness of breath; hypertrophy and dilatation of the right ventricle follow, and finally, as compensation breaks down, enlargement of the liver and dropsy supervene.

It has been stated that mitral regurgitation following upon aortic regurgitation is an element of safety to the patient, on the supposition that it tends to prevent paralysis of the left ventricle through over distension, and consequent sudden death. There is even less ground for this view in incompetence than in stenosis of the aortic orifice; for this danger will be materially diminished by the fact that the reflux from the arterial system leaves the vessels empty and lax, so that at the beginning of systole, when the chief danger exists, the strain on the ventricle is not excessive. The real danger arising from distension of the ventricle during diastole by the reflux, which incompetence of the mitral valve can in no way obviate, as blood at that time is flowing through the valve into the ventricle. If, indeed, a series of cases could be brought forward in which relief to symptoms had followed the supervention of mitral incompetence, the hypothesis might be accepted on the strength of the facts; but no such evidence is known to me, and my own observation has shown that the onset of mitral regurgitation does not prevent sudden death or afford relief, but is a further element of danger.

Prognosis.

When, with a diastolic murmur, the aortic second sound is distinctly audible in the neck, the pulse exhibits the collapsing character only in a moderate degree, and the dilatation and hypertrophy of the heart are inconsiderable; that is, when the physical signs indicate that the lesion is slight in extent, the patient may enjoy life and do hard work untroubled by symptoms for many years, provided that the lesion is due to acute endocarditis and not to degenerative changes.

This may be illustrated by an example. A medical man, aged forty, called on me in October, 1883, whom I had known and examined sixteen or eighteen years previously, when he was under Dr. Sibson's care as a student at the hospital. He had had aortic incompetence ever since an attack of acute rheumatism at the age of fourteen, so that its duration was twenty-six years. The following is an account of the physical signs which were identical with those present when he was at the hospital: "Apex beat in sixth space not much below and very little to the left of the normal situation, being a good push, but not violent or extensive; the first sound good, the second sound also distinct. A systolic murmur, probably conducted from the aorta, is audible at the apex. At the right second intercostal space a weak systolic murmur is audible, and along the right edge of the sternum, from the third cartilage downwards, over the lower part of the sternum and over the left costal cartilages, a very distinct, long, smooth diastolic murmur is heard. There is a good second sound, audible at the right second space and sterno-clavicular articulation and in the neck. The pulse is short and is felt in the fingers as well as at the wrist, but is scarcely collapsing; the carotid pulsation is marked, but the pulse at the wrist and in the temporals is invisible." He had gone through his medical studies with distinction, and had ever since carried on a hard-working country practice, with all its incidents of night-work and exposure and excessive fatigue. He could walk uphill or run upstairs without experiencing any inconvenience, and it was not for any cardiac symptoms that he consulted me in 1883.

Aortic incompetence sufficient to abolish the second sound in the neck, and to give rise to considerable dilatation and hypertrophy and to well-marked collapsing pulse is serious, though it may for many years be compatible with apparent health and strength. The patient will be capable of ordinary work and exercise, but will be sooner out of breath on going uphill or from any unusual exertion than a man in health, or may feel suddenly faint and giddy instead of losing his breath; emotion again will more readily induce palpitation. The future in such circumstances must be estimated on the principles laid down in the chapter on general prognosis. After calculation of the degree of valvular inefficiency, we must consider the tenacity of the family constitution and the vigour of the individual, as well as his age, habits, occupation, and circumstances, bearing in mind always that even in the most favourable cases the nutrition of the enlarged heart will not be well maintained after middle age, and that as years go on there is a tendency to increased resistance in the peripheral circulation. We must look, therefore, for failure of the heart and accession of symptoms sooner or later, at best long before the natural term of life. There is also the possibility that the compensatory equilibrium may at any time be dangerously disturbed or finally overthrown by imprudent exertion or anxiety or acute illness of any kind.

The **Age** of the patient at the time when the lesion is acquired is a most important consideration in prognosis, and a case may be quoted to show how long a severe lesion may be survived even under unfavourable circumstances, when the patient is young, and the heart can take on compensatory changes. The patient, a boy of fifteen, came under my observation in 1868, when he was admitted to St. Mary's Hospital, complaining of shortness of breath. His condition on admission was as follows: "Heart impulse extensive and violent, apex beat in the fifth and sixth spaces just outside the nipple line. Loud systolic and diastolic murmurs, audible over all the cardiac area, especially at the lower end and down the right border of the sternum. Pulse large, sudden, and collapsing; no second sound audible in the neck." The patient was a greengrocer, and continued to do his work, but frequently attended the hospital as an out-patient. Five years later, a mitral systolic murmur first developed, and by that time the heart was of great size; the apex beat was in the seventh space in the axillary line, and was a forcible thrust, which could be seen through his clothes. Dilatation and elongation of the aorta had carried the outer side of the ascending arch beyond the right edge of the sternum, so that, pressed into the second, third, and fourth spaces, the fingers came upon marked pulsation and thrill. He was still able to do his work, and a year later at the age of twenty-one, he married and set up in business for himself. He had now to go to the early market regularly instead of occasionally, and had more work. This he did pretty well for a time, but he soon had to spare himself, and his business fell off. In 1875, seven years from the time he was first seen, he was ill at home with rheumatism, and then lived in a basement. In 1880, when again attacked by rheumatism, he consented to enter the hospital as in-patient, which he had refused to do previously. He was again in the hospital for sub-acute rheumatism from March 5 to May 8, 1882. The rheumatism quickly subsided, and there was no pain after March 18; he had, however, a severe cough. During his convalescence, the kind of irregularity of pulse and heart beat characteristic of cardiac failure in aortic regurgitation was manifest; after three or four regular and fairly equal beats, a weak supplementary beat would follow too quickly, or while perceptible at the heart would be missing at the wrist, or the irregularity might go further than this; sometimes also the pulse would reveal two distinct efforts of the ventricle in its systole, and have the "bis feriens" character. Occasionally he had a severe attack of dyspnœa. He improved, however, and after being up and about the wards for some time, was sent to a convalescent hospital, where after doing well and gaining strength, he died suddenly, fourteen years after he was first seen. A post-mortem examination revealed old-standing extensive disease of the valves, which had in effect ceased to exist as valves, great general dilatation of the aorta, some thickening of the mitral flaps and shortening of the tendinous cords, extreme dilatation and hypertrophy of the left ventricle, some dilatation and hypertrophy of the right ventricle without any affection of the valves of this side of the heart. The weight of the heart was forty-two and a half ounces.

But even in young subjects, as soon as marked symptoms begin to present themselves, danger is at hand; when they are not habitually present, the readiness with which they are induced by exertion serves as a dynamic test of great prognostic value. Some years since, I examined, within a short time of each other, two boys with very extensive aortic incompetence, attended with almost the maximum degree of dilatation and hypertrophy. The physical signs could only tell us that there was great incompetence, and great compensatory change in both, but in the one the heart was readily put off its balance, and serious symptoms were induced by slight exertion; in the other, this was not so. The prognosis was, therefore, widely different in the two cases. The former died suddenly three years after I first saw him, while the latter did not die till ten years later.

In aortic regurgitation, acquired late in life, the prognosis is rarely favourable. Even if the lesion be the result of acute endocarditis, and therefore stationary, the heart is unable to undergo adequate hypertrophy, and efficient compensation will not be established. But after middle age acute endocarditis is of rare occurrence, and the incompetence will usually be due to degenerative changes in the valves; the lesion will therefore be progressive, more especially if the arterial tension, the primary cause in all probability of the trouble, is not carefully kept down by suitable treatment.

Prognosis in Aortic Regurgitation with Stenosis.

When aortic stenosis gradually becomes established in a case where aortic incompetence already exists of such severity that symptoms are readily and easily induced, it may act to a certain degree as a curative agent, by limiting the amount of regurgitation and making the circulation more equable and regular. As the stenosis makes its effects felt, the pulse will lose to a great extent its collapsing character, and the artery will not empty so completely between the beats. The heart will undergo further hypertrophy, and the risk of undue dilatation will be diminished. Dyspnœa will be less readily induced, and the patient will be less liable to syncopal attacks.

I have seen several cases in which a patient, after being in imminent danger from aortic regurgitation, has, on the supervention of aortic stenosis, been enabled to enjoy a life of comparative comfort for many years. It does not follow that the onset of aortic stenosis is of favourable prognostic import in all cases of incompetence; it is only when the aortic valves are severely damaged and are incapable of checking the regurgitant stream in any effectual degree, that this holds good, and when the patient is still young enough for further cardiac hypertrophy to take place.

Treatment.

Perhaps more can be done to prolong life and postpone suffering in aortic regurgitation than in any other form of valvular disease; at any rate, it is in this disease that the greatest difference can be made by care on the one hand and imprudence on the other. A patient may die suddenly from a single rash act who might have lived twenty years, or condemn himself by a single imprudence to a short and suffering existence when fair health and many years of life were possible for him. It is more especially shortly after the occurrence of the lesion, before full compensatory hypertrophy has had time to take place, that such accidents are likely to happen. Hence it is especially important when aortic regurgitation has been recently established in acute rheumatism that prolonged rest should be insisted upon.

Six or eight weeks in bed, and after this rest in the recumbent posture for another month or six weeks is advisable, and the boy should not go to school, or the young man to business for another six, eight, or twelve months, according to circumstances. It is even possible that a diastolic murmur may completely disappear, and with it all symptoms of aortic incompetence, when due care of this kind has been taken.

Not uncommonly incompetence of the aortic valves is discovered unexpectedly at an interval after an attack of rheumatism during which endocarditis had not been suspected, or had not been revealed by murmurs. It is a most important and useful precaution, therefore, to examine the heart at intervals for some time after rheumatic fever, and it is an imperative duty to do so when there has been cardiac complication of any kind or degree.

In children, as the articular manifestations of rheumatism are usually so slight, and the heart is so frequently attacked, an examination of this organ should never be omitted; there may be nothing to suggest the presence of endocarditis; but a history of transient pains in the joints with febrile disturbance should at once arouse suspicion, and the presence of rheumatic nodules is almost pathognomonic of cardiac mischief present or to come. Irreparable damage to the heart not infrequently results from a child being allowed to go about with unsuspected endocarditis.

Caution and rest are necessary, not only after an attack in which there has been endocarditis setting up aortic regurgitation, but also after a febrile attack of any kind complicating cases of aortic incompetence. Dilatation is easily induced in a heart weakened by fever under the continual strain to which the left ventricle is exposed, and there is special liability to sudden death during the period of convalescence from acute illness.

On the other hand, provided care and complete rest are insisted on, the heart may actually regain lost ground from the diminution of resistence in peripheral circulation due to the arterio-capillary relaxation attending pyrexia. In a severe case of aortic regurgitation, I have known an old-standing mitral regurgitation from secondary dilatation of the left ventricle, disappear for some time under these circumstances.

The risk of sudden death makes it imperative that the patient suffering from aortic disease should be specially warned against over-exertion and hurry, such as running to catch a train or running upstairs, or excessive mental excitement of any kind. The fatal event does not always take place during the exertion or excitement, but may be postponed till the next day. Periods of rest from time to time may be of striking service. Whenever exaggeration of the short and sudden character of the pulse is observed, especially if there is a faltering in the beat now and then, and still more when the apex of the heart is found to be receding outwards, and the beat is becoming diffuse, the recumbent posture should be enforced for some two to six weeks.

As compensation fails and symptoms become more continuous and severe, they will arise from and take their characteristics from one of two causes : either (1) failure on the part of the left ventricle to maintain a sufficient movement of blood in the capillaries, or (2) backward pressure in the veins, the result of right ventricle failure secondary to that of the left. In the former case, sudden death from syncope with little warning or apparent cause is liable to occur; failing that, there may be sleeplessness of a peculiarly harassing kind, or painful dyspnœa, unexplained by any interference with the entry of air into the lungs, or by want of aëration of the blood. The face will be pale and have an anxious and suffering expression ; the patient will be very restless and weary. Dropsy, if present, will be slight, though there may be some fluid in the pleural cavities with œdema of the bases of the lungs.

Under these circumstances the object of treatment is to

sustain the failing heart by nourishment, stimulants, and such remedies as may contribute to this end—nux vomica or strychnia, with ammonia and ether, to which belladonna or atropin often makes a valuable addition; sometimes digitalis may be of service, but it must be used with caution; or strophanthus, which theoretically ought to be safer as having less contractile influence on the arterioles. Morphia hypodermically is often of the greatest possible benefit, giving quiet sleep, which is not only an inexpressible comfort to the sufferer, but frequently recruits the strength, and, by rendering the recumbent posture possible, so far relieves the heart that it proves the starting-point for a temporary recovery. Occasionally, however, a patient, after a good night's rest procured in this way, will say he feels better, sit up in bed, and suddenly fall back dead. This danger of syncopal attacks should be explained to patients and friends, so that watchfulness and care on their part may guard against any imprudent effort or sudden movement which might be attended with such fatal consequences. Morphia or opium administered by the mouth is much less effectual, bromides are mostly useless, chloral is positively dangerous.

Angina pectoris may complicate this form of heart failure. It is usually relieved by nitrite of amyl or nitroglycerine, and patients suffering from anginoid pain complicating aortic regurgitation may come to take the latter remedy in extraordinary doses. When relief cannot be obtained in any other way, morphia may be given hypodermically. Whatever tends to strengthen the heart or relieve it from work will tend to prevent the onset of anginoid symptoms, and when such threaten, treatment must be directed to these two points.

Arsenic, and more especially phosphorus, have had in my hands a very beneficial influence as cardiac tonics in such cases.

When the preponderating character of the symptoms is that of venous obstruction, the jugular veins being distended and pulsating, the liver enlarged, and dropsy present; when, in fact, we have with aortic physical signs mitral symptoms, the line of treatment is quite different to that just described.

Purgatives should be given in full doses, but judgment and caution must be exercised in their administration, as unfavourable effects may develop abruptly. Digitalis may then be given, usually with great benefit, and when there is much dropsy, diuretics in addition to the digitalis will often have a marked effect. It is under such conditions that digitalis finds its opportunity in aortic regurgitation, and justifies the statements of those who find this remedy of the same service in aortic as in mitral disease.

If the administration of digitalis is persisted in after the recovery from dropsy and the more severe symptoms of venous stasis, it is not uncommon for patients to die suddenly, sometimes before leaving bed, more frequently when they have begun to get up and move about. There are grounds for suspicion that digitalis contributes to this, but sudden death may occur whether it has been left off or not. In the absence of mitral symptoms, it is rarely that digitalis is called for in aortic incompetence or is of service, and it may undoubtedly do harm. It may set up sickness, which is an ominous symptom in this form of disease, and induces a condition of asthenia difficult to remedy.

Venesection, even when the venous stasis is severe, is rarely to be contemplated. It is true that the relief to the right ventricle so afforded might enable it to regain control over its contents and increase the supply of blood to the left ventricle and the amount available for distribution to the arterial system, but before this occurs there may be a momentary faintness, leading to a fatal attack of syncope.

CHAPTER XII.

MITRAL INCOMPETENCE OR REGURGITATION.

PHYSICAL SIGNS—THE MURMUR OF MITRAL INCOMPETENCE—THE PULSE—EXPLANATION OF IRREGULARITY OF PULSE—MITRAL INCOMPETENCE DUE TO ACTUAL CHANGE IN VALVES, THE RESULT OF ENDOCARDITIS—ESTIMATION OF EXTENT OF LESION: (1) FROM CHARACTER OF MURMUR AND FIRST SOUND; (2) FROM COMPENSATORY CHANGES IN THE HEART; (3) FROM SYMPTOMS—PROGNOSIS—MITRAL INCOMPETENCE WITHOUT DAMAGE TO VALVES - ITS CAUSATION AND EXPLANATION—DIFFERENTIAL DIAGNOSIS—THE MITRAL INCOMPETENCE OF MIDDLE OR OLD AGE—TREATMENT.

MITRAL INCOMPETENCE OR REGURGITATION.

Physical Signs.—The evidence of regurgitation through the mitral orifice is a systolic murmur heard in the region of the apex, and very frequently beyond the apex in the fifth or sixth space towards the axilla. It is often audible also at the back of the chest, between the scapula and the spine, at about the middle of the posterior border of the scapula. The murmur reaches the surface at this point, not by extension round the thorax from the axilla, but by an independent route, being conducted by the vertebræ from the base of the ventricle, the shoulder of which rests upon the spinal column. Occasionally the murmur of mitral regurgitation is heard more loudly in the fourth or even in the third space in the vertical nipple line or just outside it, than at the apex. When this is the case, an impulse is usually to be felt at or near the same point.

Murmurs which may be mistaken for Mitral Regurgitant Murmur.

The only murmurs likely to be mistaken for this are, a systolic aortic murmur conducted to the apex, a systolic and therefore regurgitant tricuspid murmur, and the spurious murmur produced by compression of the edge of the lung by the ventricular systole. A pulmonic systolic murmur is also occasionally heard almost at the apex. A question which frequently arises when a systolic murmur is heard both at the base and at the apex, is, whether it is one and the same, in which case it will be aortic, or whether it arises from two independent sources. It is only when the murmur is loudest over the aorta that there can be any uncertainty, and as a rule it is not difficult to decide. Any marked difference in the tone or character of the apex murmur, as compared with that heard at the base, or a superadded musical element in it, would be sufficient to show that it was not the conducted aortic murmur, and where such evidence as this is lacking, if the apex murmur is longer than the aortic, and especially if it is conducted outwards for any distance, or is heard in the left interscapular space, there can be no doubt. It must be remembered, however, that a loud, rough, aortic murmur may be heard as a smooth murmur of a different tone at the apex, the thick muscular wall of the ventricle not lending itself to the propagation of coarse vibrations, which are, in fact, vibrations of the aorta itself. A loud aortic murmur, again, may be audible all down the thoracic spine.

With regard to the tricuspid murmur a special warning

may be of service, as it is not unfrequently set down as mitral. The tricuspid area, so called, is over the lower costal cartilages, near the ensiform appendage, but a systolic tricuspid murmur is often heard as far out as the apex, and occasionally has its maximum intensity just to the inner side of the apex. When a murmur heard at the apex is lost immediately to the left of the beat, while it is audible between the apex and the lower end of the sternum, its seat of production is at the tricuspid, and not at the mitral orifice.

An imitation of a systolic apex murmur is not unfrequently produced by movement of air in the lungs overlapping the heart as it is compressed by the systole of the heart, and it is sometimes loud enough to be heard by the subject. The imposture is easily detected, by the fact that it is only audible, as a rule, during inspiration, or is, at any rate, much more distinct then, and that its distribution does not coincide with the conduction of the true mitral murmur, but follows the line of the thin edge of the lung across the pericardium from the neighbourhood of the apex towards the sternum.

The sources of error just mentioned being eliminated, a systolic murmur audible at and to the left of the apex, is mitral in origin, and indicates incompetence in the valve and reflux of blood into the left auricle. There is not the same doubt as to its significance as there is with regard to the constant association of obstruction with an aortic systolic murmur; a mitral systolic murmur means regurgitation. It does not, however, necessarily imply that there is actual disease of the mitral valve, as there are other causes, enumerated later, which may give rise to mitral regurgitation from dilatation of the left ventricle or stretching of the mitral orifice.

The Pulse of Mitral Regurgitation.

The distinguishing characteristic is irregularity, both in rhythm and in force. In advanced cases no two beats are alike; a few fairly strong, full pulsations, at something like proper intervals, will be followed by a number of feeble, hurried beats, these again perhaps by a single good stroke, after which there is a pause and then renewed hurried action. Little difference of opinion exists as to the association of mitral incompetence with irregularity of pulse. It would not be true to assert that there can be no considerable regurgitation without irregular action of the heart, but it is a safe working hypothesis that a mitral murmur is not attended with serious reflux while the pulse remains regular. What, then, is the cause of the irregularity? We find similar irregularity occasionally attending dilatation of the heart, and it might be suggested that, as mitral insufficiency is not unfrequently the result of dilatation of the ventricle, the irregularity of the pulse is symptomatic of the dilatation, and not of the regurgitation; but the irregularity accompanies incompetence caused by rheumatic damage of the valves, quite as constantly as it does the insufficiency due to dilatation.

Irregular heart action and pulse, again, may be among the final symptoms in any form of valvular disease, though they are rarely quite of the same character as in mitral regurgitation; but in this latter disease the irregularity supervenes early, and is not inconsistent with fair compensation and apparent health for many years.

It is probable that the varying pressure to which the heart is subjected in inspiration and expiration may account for it. The variations of intrathoracic pressure in ordinary respiration have no obvious effect on the action of the heart; but we can at any time slightly disturb the rhythm of the heart and pulse by taking and holding a very deep breath. When again there is incipient irregularity careful observation will almost always show that the break in the rhythm occurs at the moment when there is a change in the intrathoracic pressure at the beginning and end of inspiration or expiration. This is best seen, perhaps, in cases of bronchitis and emphysema with dilated right heart. The thin-walled auricles will be all the more susceptible of this change of pressure when distended and dilated, while it will scarcely affect the ventricles.

The heart works habitually under a minus pressure due to the traction on the pericardium by the elasticity of the lungs, and is only subjected to pressure, strictly speaking, under normal conditions when the chest is filled and the glottis closed in effort. The aspiration resulting from this minus pressure normally aids the circulation by drawing the blood into the auricles, which then carry it on into the ventricles.

When the left auriculo-ventricular valves are incompetent, the pressure on the auricle during expiration will aid in resisting the systolic reflux from the ventricle, and in driving the blood into the ventricle during diastole ; at the moment when the positive pressure of expiration is exchanged for the negative pressure of inspiration, the resistance to the reflux of blood into the auricle is suddenly diminished, so that more will be allowed to regurgitate, and less will be carried into the aorta ; there will be also less assistance from the pressure upon the auricle in filling the ventricle. Corresponding differences of an opposite kind will attend the end of inspiration. The varying supply of blood to the ventricle thus induced, cannot fail to produce a tendency to irregularity.

MITRAL INCOMPETENCE DUE TO LESIONS OF THE VALVES.

The causation of mitral incompetence is most varied. It may be due to actual damage to the mitral valve by acute or chronic endocarditis, or to imperfect apposition of the valves or stretching of the orifice from dilatation of the left ventricle. The former group, where the valves have been injured by an inflammatory process, will be first considered.

The kind and degree of deformity produced by endocarditis may vary greatly. The edges may be thickened and shrunken, presenting, instead of the translucent, thin, crenated margin, an opaque, even, rounded border, in which the delicate marginal ramifications of the tendinous cords are swallowed up and lost. A considerable area of valve may still be available for closing the orifice, although the apposition may not be accurate. In a more advanced state of change, the body of the cusps may be opaque, thickened, and contracted so that a considerable gap must remain, allowing of great reflux even when their approximation is greatest; and sometimes this is carried so far that, as valves, they are practically non-existent, especially when the chordæ tendineæ are greatly shortened, which they may be to an extent which brings the margins of the flaps close to the apices of the papillary muscles.

Chronic endocarditis or degenerative change may produce great deformity and insufficiency, but generally there is more rigidity and less shrinking than after acute endocarditis, while calcification is not uncommon.

ESTIMATION OF THE EXTENT OF THE LESION.

Our first inquiry will be as to the physical signs and symptoms by means of which we may approximately infer the degree of impairment of the valves, and the amount of regurgitation which takes place.

The murmur affords, in some cases, important information. When it is not conducted much beyond the situation of the apex beat, the apex itself not being greatly displaced, and is not audible in the back, the regurgitation is usually slight. But an exception to this statement may be found at the opposite extreme, when the orifice is gaping and the ventricle weak. Under such circumstances the murmur may be short and scarcely audible; symptoms, however, will at once distinguish between the two conditions.

The persistence of the left ventricle first sound is evidence that there is no advanced change in the valves, more especially when the murmur is retarded. This is an important point to note. If the regurgitation is considerable, the murmur takes the place of the first sound, and is said to hide it; but it is not a mere question of loudness, there is a true substitution of the murmur for the sound. When, therefore, the first sound is distinctly heard heading the murmur, some difference exists. It would be going too far to say, as in the case of the aortic valves, that the sound indicates sufficient check to the refluent blood to produce tension of the valves and cords, and of the ventricular wall, but the inference tends in that direction. With regard to the retarded-systolic mitral murmur, when the murmur follows the first sound at a distinct though very brief interval, it would seem that there is, in the first instance, complete closure of the valves, but that during the contraction of the ventricle, the apposition is deranged so that leakage occurs. Such leakage can scarcely be other than slight. These inferences are confirmed by the fact that the persistent first sound and the retarded-systolic murmur are most common in the curable or temporary regurgitation of anæmia, and the conclusions based on this observation may be extended to other cases than those of early life. One caution is necessary, that is, to beware of taking the modified first sound of mitral stenosis, when there is both obstruction and regurgitation, for the normal first sound.

The general statement made in the first chapter that a loud and long murmur is usually significant of a lesser degree of structural damage and functional imperfection than a short and weak murmur, is another point to be borne in mind. Loudness implies strength of contraction, and length shows that the ventricle goes through with its systole, and also that it is not quickly emptied, as it would be were there a large escape into the auricle. The murmur of worst significance is a short, weak whiff, varying in intensity and duration in successive beats.

It is the mitral systolic murmur which is most frequently musical; very commonly, but not perhaps always, the musical murmur indicates a narrow chink, and therefore little reflux. The musical note may be high-pitched or low, nearly always it is accompanied by a blowing murmur, and, according to my experience, has never the pure tone sometimes produced at the aortic orifice. It does not always begin at the same time with the blowing murmur, but may be interpolated into it, beginning later and ending sooner.

As the action of the heart is very frequently irregular, the intensity, character, pitch, and length of the murmur, may vary greatly from one beat to another.

For further evidence we must trace the effects of regurgitation through the mitral orifice. The immediate effect of reflux of blood into the left auricle will be to distend this cavity, but, lying deeply as it does, the physical signs of an early stage of dilatation are not such as can be relied upon for definite information; but another result which follows early is obstruction to the free outflow of blood from the pulmonary veins into the auricle, and this soon makes itself felt backwards through the capillaries of the lungs, and gives rise to high pressure in the pulmonary artery. Here we come upon audible evidence; the increased tension brings down the semilunar valves with greater force, and gives rise to accentuation of the pulmonic second sound. It is not easy to say what degree of intensification of this second sound is required as proof of obstruction to the flow of blood through the lungs. The aortic second sound has to be taken as a basis for comparison, and while most observers say that the aortic is the louder of the two, my own conclusion is that the pulmonic is usually louder than the aortic. Again, the pulmonic second sound may be accentuated by resistance arising in the lungs themselves, as in bronchitis and emphysema. Marked and persistent intensification of the pulmonic second sound, when a mitral systolic murmur is present, must, however, be taken as evidence of sufficient reflux to increase the pressure in the pulmonary artery by the obstruction to which it gives rise.

From this follow hypertrophy and dilatation of the right ventricle. The call upon the right ventricle for increased force in propelling the blood through the lungs results in a combination of dilatation and hypertrophy, of which the physical signs are displacement of the apex outwards or to the left and undue right ventricle impulse, felt when the hand is placed over the lower costal cartilages to the left of the sternum, and seen, and perhaps felt, in the epigastrium and below the left costal margin. The right ventricle in effect comes to the aid of the left; the heightened pressure maintained by its means in the left auricle resists the reflux of blood from the ventricle, and since life is often sustained when the valves are practically non-existent, the intra-auricular pressure must in such cases almost, if not quite, equal the pressure in the systemic arteries. Further, the same force which resists the reflux must, on the relaxation of the left ventricle, drive the blood violently into it, taking it, so to speak, at a disadvantage, and distending it in the same way, though not to the same degree, as aortic regurgitation, giving rise thus to dilatation, which is accompanied or followed by more or less hypertrophy. In proportion as the left ventricle is enlarged, the apex will be carried downwards and its beat become conspicuous, and the hypertrophy of the right ventricle pushing it outwards and sometimes upwards, the ultimate displacement of the apex will be the resultant of the changes in the two ventricles. The volume of the heart as a whole will be increased, giving rise to a corresponding extension of the area of deep dulness.

These changes, the necessary result of the incompetence of the mitral valve, traceable directly to the mechanical difficulty produced by it, become the measure of the difficulty and of the incompetence. A systolic mitral murmur without increase in the area of cardiac dulness, without much displacement of the apex, and without marked accentuation of the pulmonic second sound, without, again, symptoms indicative of disturbance of the circulation such as might be due to failure in the compensatory changes, is not attended with any considerable regurgitation, and is not a source of present danger. Unless there is reason to look upon it as the beginning of progressive damage, it may practically be disregarded. As the signs of pressure in the pulmonary circulation and of changes in the walls and cavities of the heart increase, we infer increase in the amount of reflux, for they indicate greater difficulty in neutralizing its injurious effects. We, therefore, calculate on diminished stability of the compensatory balance and on less power of regaining a working equilibrium of the circulatory forces if it is once overthrown.

To sum up the indications by which the extent of the lesion may be surmised:

The Pulse.—While it may safely be said that if the pulse remains regular the reflux is slight, the degree of irregularity of the pulse will not be any criterion as to the extent of the reflux. More is to be learnt from the degree of vigour of the beat, and the length of the pulse wave.

Further information will be derived from the character of the murmur, and the extent to which it replaces the first sound, and again from accentuation of the pulmonic second sound; but the amount of dilatation and hypertrophy of the heart, more especially of the right ventricle, will afford a more accurate basis for the estimation of the degree of incompetence.

The symptoms must also be concurrently taken into consideration: they will when present be such as have already been described in Chapter IV.; it is, therefore, not necessary to enumerate them again here. No symptoms will be present in cases of slight regurgitation, unless there is anæmia or some other complication. In severe cases the size of the liver will afford important information as to whether the hypertrophy of the right ventricle has enabled it to cope efficiently with the effects of the mitral lesion; the degree of enlargement of the liver, therefore, together with the degree of readiness with which other symptoms such as dyspnœa and dropsy present themselves, will afford indirect evidence as to the extent of the lesion in the absence of degenerative changes in the heart or of other complications. When the symptoms of cardiac embarrassment are more severe than would be expected from the extent of the lesion as estimated by the physical signs, adherent pericardium, or degenerative changes in the heart, must be borne in mind as a possible cause of this.

Prognosis.

The range of possibilities as regards duration of life in mitral regurgitation, due to actual damage of the valve by acute endocarditis, is more extensive than in any other form of valvular disease. It is the least serious and the most amenable to treatment of all the valvular affections.

When the lesion is slight the patient may live to old age without experiencing inconvenience from the results of the lesion, and may be capable of much hard work and pass through serious illnesses. In moderately severe cases, with reasonable precautions, he may live many years without serious symptoms declaring themselves. Women with mitral regurgitation may bear children with impunity.

In endeavouring to forecast the further course of a case of mitral incompetence when the regurgitation is inconsiderable, or when compensation has been established, one of the most important considerations will be the degree of arterial tension. Everything, so to speak, will depend upon the amount of resistance in the peripheral systemic circulation, which may vary extremely. If with a given degree of mitral incompetence, we have undue arterial tension, the force of the regurgitant stream will be great, the back pressure in the auricle and pulmonary circulation high, the demand upon the right ventricle severe, all of which are circumstances tending to the production of structural changes; if, on the other hand, the arterial tension is low, the conditions are reversed.

The case of a patient, whom I have known to have mitral regurgitation for at least thirty-five years, may illustrate this. When first seen there were disquieting symptoms, a sense of constriction and oppression at the chest on moderate exercise, and liability to slight fainting attacks. Finding with such symptoms the pulse extremely short, soft, and compressible, it was at first feared that there were degenerative changes in the heart, but the family pulse, as exhibited by several children and grand-children, was one of extremely low tension, and this was found to be the explanation of the patient's weak pulse. There can be no doubt that the absence of resistance in the systemic arterioles and capillaries had an important share in the prolonged immunity from structural changes which he has enjoyed in spite of work of the most trying and arduous kind.

One or two other examples of prolonged survival may be mentioned. Two gentlemen, in whom I found mitral regurgitation more than twenty years since, were, when I last saw them, more than forty years of age and in active work. The murmur was loud and long in both, and in one there was considerable hypertrophy of the right ventricle and marked irregularity of the pulse. Another gentleman, whom I have repeatedly examined, and found to have a mitral systolic murmur, was, thirty-five years before I saw him, condemned to life-long inactivity, and I learnt from the family medical man that there had been a continuous history of valvular disease for that length of time; he rebelled against the sentence, and was still at the age of sixty-four doing strenuous work. Nothing is more common than to find a mitral systolic murmur after the age of seventy, but we cannot say how long it has been present, as it is not often that we can date it. Mitral regurgitation is the disease which sends patients into hospital time after time, often with advanced dropsy or severe pulmonary complication, from which they recover so completely as to resume work for a while. It is again the affection most commonly met with in out-patient hospital practice, while among in-patients and in the post-mortem room it is less frequently represented than mitral stenosis. All these facts point to one conclusion, viz. that mitral incompetence is not a deadly form of heart disease; I consider it to be less serious than aortic stenosis. It is not merely that in a large number of cases the actual damage to the valve and the consequent functional imperfection is only slight, but very extensive change, provided it is stationary, may be survived twenty or thirty years, and not prevent the patient reaching old age in comfort.

Mitral Incompetence without Damage to Valves.

In the large group of cases of mitral regurgitation now to be considered, the cause of the reflux is not actual damage to the mitral valves, but stretching of the orifice or imperfect closure of the valves, due to dilatation of the left ventricle.

Such dilatation may be secondary to aortic disease, or the result of prolonged high arterial tension; but mitral regurgitation due to these causes need not be considered here, as its prognosis and treatment are those of the primary affection, and have been discussed elsewhere.

The chief causes of mitral incompetence in the group to be considered, are: anaemia, acute febrile disorders, and conditions associated with advancing age.

The so-called haemic mitral murmurs which have sometimes been supposed to have some other significance than actual reflux, as they disappear with improving health, are not unfrequently met with in anaemia, especially in young women, or at times after acute febrile diseases, such as acute rheumatism, typhoid fever, or measles; the systolic mitral murmurs so commonly present in chorea sometimes belong to the same class, and mitral regurgitation, coming on late in life, not unfrequently has a similar causation, although it may not be removed by treatment.

It is to this class of cases that McAlister's [*] explanation of mitral reflux without disease of the valves or enlargement of the orifice, already previously referred to, applies. It had been forgotten or overlooked, notwithstanding ocular demonstration on the living heart, and preparations showing the heart in systole and diastole by Sibson, notwithstanding also measurement of the circumference of the heart in the contracted and relaxed condition and the fact that the opening was only the upper end of the ventricular cavity, that contraction of the orifice contributed in an important degree to its perfect closure by the valves. When regurgitation occurred, which was not explained by changes in the valves, it was accounted for by

---

[*] *British Medical Journal*, August, 1882.

irregular action of the papillary muscles, or by the supposition that the papillary muscles, and with them the tendinous cords, were carried by dilatation of the ventricle to such a distance from the base of the valves that the margins, being dragged down by the chordæ tendineæ, could not meet. It was demonstrated in McAlister's paper, that all that is required in order that the reflux in question may occur, is that from languid contraction of the cardiac muscular fibres, or from resistance in the peripheral circulation, the due constriction of the orifice should not take place; the valvular mechanism will then be deranged and the apposition of the flaps be incomplete. It must be added, however, that if the contraction of the orifice is imperfect, so will be the ventricular systole as a whole, so that the position of the papillary muscles may contribute to the derangement.

The Mitral Murmur of Anæmia and Debility.

This, though only a particular case of regurgitation from dilatation, is worthy of separate consideration. It is met with in only a small proportion of cases of anæmia, and may be absent when the bloodlessness is extreme. It does not, therefore, mark a certain degree of impoverishment of the blood, and may indeed be found when this is by no means advanced. The remarks apply equally to chlorotic anæmia, to anæmia from loss of blood, and to pernicious anæmia. The fact that a mitral murmur is so often absent in fatal pernicious anæmia is perhaps the most remarkable. It seems probable that while the poverty of the blood and the consequent innutrition of the heart are the predisposing influences, the proximate cause is some overstrain, which may be sufficient even in the absence of anæmia.

As an instance of this, may be quoted the case of a young lady whom I saw from time to time during a period of four years. She was well nourished, with well developed muscles, as well as a fair amount of adipose tissue, and the lips and face were a good colour. When first seen she had a pronounced mitral murmur, and the apex beat was displaced to the left of the nipple line. There were no murmurs in the neck or over the pulmonic area or aorta; the catamenia were excessive.

This was her condition in July; in October the murmur had disappeared and the apex had receded. In the following January, after the death of a relation by drowning, which she witnessed, the murmur and dilatation had returned. A month later the murmur could not be heard, but the apex was still outside its normal situation. On examination in July, September, and December of that year, the apex beat was normal and the murmur absent, although the menses were suspended for long periods, and she had lost flesh. No further relapse had occurred when I last saw her. The influences which appeared to determine the occurrence of the regurgitation were the relaxing climate in which her home was placed; the frequent necessity, from the situation of the house, of walking uphill when tired; and, probably more than all, domestic worry.

There is, however, in anæmia, a cause of dilatation of the weakened left ventricle independent of external and accidental circumstances; this is high arterial tension, which, as has been elsewhere pointed out, nearly always attends this state of the blood. Undue arterial tension implies increased resistance to the emptying of the ventricle, and this gradually distends it. When regurgitation is not produced, there are still very frequently the signs of dilatation in displacement outwards and downwards of the apex, and a diffuse character of its beat. The operation of this cause perhaps throws light on the production of permanent valvular disease by anæmia, which Goodhart has shown to be probable. The peripheral resistance throws extra stress upon the mitral valves, the action of the heart is liable to be excited and violent in anæmic subjects, which will intensify the effects of the strain, and so valvulitis may be set up.

Differential Diagnosis.

It may sometimes be difficult to decide whether a mitral murmur is due to conditions such as have just been described, or to actual changes in the valves. Where there are obvious compensatory changes in the left or right ventricle, there can be no doubt, and in many cases the state of health of the patient with the history will be a sufficient guide; but difficulties may arise, for example, when there is marked anæmia in a woman who has had acute rheumatism, or when a murmur is present, without anæmia adequate to account for it, in a patient who is not known to have suffered from rheumatism. In such doubtful cases, the character of the murmur may afford useful information. If the murmur is soft and blowing, and a murmur is heard not only at the apex, but at the pulmonic and aortic areas, the probability is that it is due to anæmia, and not to actual valvular change.

Hæmic murmurs are usually soft and blowing in character, not harsh or musical, and they do not replace or extinguish the first sound. Generally speaking, they are not conducted to the axilla or heard over the back, and are not accompanied by much displacement of the apex beat. While this is the rule, I have known several cases in which it was departed from, the apex being displaced, and the murmur heard towards the axilla or over the back.

Furthermore, it is the hæmic murmur which is most frequently late-systolic in time, that is, it follows the first sound at an appreciable interval instead of commencing synchronously with it.

These indications, together with the history of the case and the physical signs present, will as a rule be sufficient to clear up the diagnosis in doubtful cases.

REGURGITATION IN ACUTE FEBRILE DISEASES.

Regurgitation of Similar Character resulting from Acute Disease lasts for a short time only, and is chiefly interesting from the fact that it is not uncommon after acute rheumatism, the great cause of true valvular lesions. It is often impossible to say whether a soft mitral murmur heard towards the close of an attack of rheumatic fever, indicates the beginning of actual disease, or is the result of temporary functional imperfection, and time alone can decide. If it occurs after the commencement of systole, and especially if a pulmonic systolic murmur is also present, it may be hoped that the mitral regurgitation is produced in the same way as in anæmia.

Chorea.—The mitral murmur associated with chorea is often louder and more harsh than that heard in anæmia; not unfrequently also it persists after the chorea is cured. The regurgitation may, in fact, either be the result of endocarditis which has damaged the valves during acute or sub-acute rheumatism, which is so frequently an antecedent of chorea, or it may be merely functional.

The cases considered constitute the mitral incompetence of early life, in which there is no reason to believe that the orifice is much larger than normal. But mitral incompetence may be established during the decline of life, and at this period the auriculo-ventricular opening may be greatly enlarged. This class of cases must be briefly considered.

THE MITRAL REGURGITATION OF MIDDLE AND OLD AGE.

It is astonishing how frequently this is met with, and how imperceptibly it is established. It is possible that in some cases there is no more organic valvular or structural alteration than in anæmic regurgitation; very commonly, however, there is positive and considerable enlargement of the auriculo-ventricular opening, and no definite diagnosis can be made between this condition and changes, such as thickening and contraction, slowly taking place in the valves.

The mitral orifice may be greatly stretched so as to admit three or four fingers; usually, as the size of the opening increases, the valves seem to expand *pari passu* with it, so as to be apparently capable of closing it more or less perfectly when tested by water in the usual way after death, but during life regurgitation is generally permitted as a result of imperfect co-aptation. Dilatation of the ventricle and stretching of the auriculo-ventricular opening usually take place together, but by no means to a corresponding extent; either orifice or ventricle may be disproportionately enlarged. As the regurgitation may be said to be only an incident of dilatation, a full examination into its causes will be deferred. Undue arterial tension will have a most important place among them, and I have watched in many cases the gradual supervention of a murmur upon sounds which have had the loud and sharp character produced by excessive peripheral resistance; but I have also found a mitral murmur when there has been no arterial tension, the pulse being large, soft, short, and compressible. In chronic disease of the kidneys, which can scarcely escape mention when the effects upon the heart of protracted high tension in the arteries are under consideration, mitral regurgitation is less frequent than might perhaps have been expected, and when it occurs, there is usually actual lesion of the valves. This is probably due to the fact that the hypertrophy, which is associated with the dilatation, and is usually the predominant change, renders the ventricular systole efficient in constricting the orifice.

Treatment.

Mitral incompetence is met with in such varied degree and has such varied causation, that it is necessary in discussing the treatment to differentiate between the most important varieties.

1. When there is no valvular lesion, and the regurgitation is due to imperfect systolic narrowing of the orifice as in anæmia, and occasionally after acute disease, the principal treatment will be that of the condition which has given rise to the atony of the cardiac muscle. When the heart is dilated after acute disease, a period of absolute rest may occasionally be necessary. Violent or sustained exertion in such cases and in anæmia, must be avoided, as acute dilatation may be produced, or a pre-existing dilatation may be increased by a comparatively slight cause. Gentle regular exercise will be of great importance, and may be gradually increased in duration and vigour, as the heart regains its tone.

A daily period of repose in the recumbent position should be enjoined, and a complete rest for a certain time before and after meals should be insisted on.

Climate has an extraordinary influence in this condition of heart, and the Œrtel system of graduated exercise at a considerable elevation is often of great service.

Iron, quinine and strychnine may be given for the anæmia and general weakness, and small doses of digitalis will often make a great difference in the results obtained.

The systolic apex murmur may be confidently expected to disappear in time, as a result of suitable treatment.

2. In the case of a systolic apex murmur with extremely little regurgitation, common after middle age, when not due to dilatation of the left ventricle, little treatment is required.

In most cases, instead of restriction in exercise being necessary, regular exercise will have to be ordered, and a sedentary, self-indulgent mode of life modified, as these patients often plead a weak heart as an excuse for avoiding fresh air and exercise, which they dislike.

When the incompetence is real, and has required

compensatory changes to neutralize its effects, it is still desirable that the patient should take regular exercise, the duration and vigour of which must be determined by his strength and breath. So long as respiratory distress is not induced, outdoor exercise will do good and not harm. Standing about indoors is much more likely to be injurious.

Bronchitis.—The one contingency to be specially guarded against is bronchitis, or any acute affection of the lungs. The pulmonary circulation is already carried on under difficulties, and superadded obstruction may intercept the compensatory high pressure maintained in the left auricle by the right ventricle; the result being that there is less opposition to the mitral regurgitant stream, the left ventricle is imperfectly filled, and the systemic blood supply is impaired: further, the right ventricle, unable to contend with the double obstacle to the transmission of blood through the lungs, becomes over-distended and crippled, so that tricuspid regurgitation and venous stasis result. The dreaded ulterior effects of mitral regurgitation are thus anticipated, and, besides this, recovery from the bronchitis is retarded by the congestion in the pulmonary capillaries.

High Arterial Tension.—Precautions must also be taken against high arterial tension, which may be due to renal disease or gout, or may be produced by too nitrogenized a system of diet, or by habitual consumption of beer or strong wines, or by constipation. The result of high arterial tension will be to increase the peripheral resistance, and thus increase the amount and force of the mitral reflux, causing an additional continuous strain on the compensatory mechanism. Suitable dieting and the habitual use of a mild mercurial purge once or twice a week, will tend to keep down high arterial tension when it is present.

TREATMENT WHERE SYMPTOMS ARE PRESENT.

When symptoms set in, and especially if any exciting cause can be traced, there is more chance of making head against them, and of improvement under treatment, than in other forms of valvular disease. The result of mitral regurgitation is always backward pressure, taking effect first on the lungs, then on the right ventricle, and finally giving rise to obstruction to the systemic venous return, when the right ventricle breaks down. Tricuspid regurgitation does not appear to add appreciably to the obstruction, and, indeed, is thought by some to prevent paralysis of the right ventricle from over-distension. The two problems in the treatment are the relief of the venous stagnation and the strengthening and restoration of tone and vigour to the right ventricle, so that it can again perform its work efficiently. There is no constricting barrier opposing a fixed mechanical obstruction to the blood current as in mitral stenosis; the task is therefore easier.

For the **Relief of the Venous Obstruction,** purgatives, of which mercury in some form is a constituent, will usually be sufficient, repeated according to their effect, and according to the condition and strength of the patient every second or third day. The application of leeches over the liver, if it is enlarged and tender, will often be of great service, and almost invariably affords relief. Venesection is not often absolutely required, though probably it might more frequently be resorted to with advantage than is the case.

Concurrently digitalis should be given, and it is in the treatment of dropsy and advanced conditions of mitral incompetence, that it may be administered with the greatest confidence and least apprehension of its so-called cumulative effects. It increases, it is true, the peripheral resistance, but as long as the structures of the heart are sound, it appears to increase the vigour of its contraction in greater proportion; this is especially the case with the right ventricle, and the improvement in the transit of blood through the lungs is perhaps the most important element in its beneficial results. It may be given with satisfactory results for any length of time, as there is no barrier in the shape of a stenosed orifice, against which the increased energy expended by the heart will be exhausted: the only reason for discontinuing it will be nausea or loss of appetite, to which it sometimes gives rise, or marked diminution in the amount of urine excreted.

CHAPTER XIII.

MITRAL STENOSIS.

ITS PREDOMINANCE IN THE FEMALE SEX—MORBID ANATOMY AND PHYSIOLOGY OF CONSTRICTION OF THE MITRAL ORIFICE—THE PHYSICAL SIGNS—THE PULSE—THE CHANGES IN THE HEART—THE CARDIAC MURMURS—THREE STAGES IN THE PROGRESS OF THE DISEASE AS DEFINED BY AUSCULTATORY SIGNS—THE CHARACTERISTICS OF THESE THREE STAGES—SYMPTOMS—DIAGNOSIS—PROGNOSIS—TREATMENT.

Mitral Stenosis.

CONSTRICTION of the mitral orifice is on many grounds the most interesting of the valvular affections of the heart. It is a common, and at the same time a serious form of valvular disease, and its clinical history presents peculiarities, some of which have long been recognized, while others have not received adequate notice. It was the last of the valvular diseases to be associated with distinctive physical signs, and to Gairdner the credit of this discovery is due. It also presents greater difficulties in diagnosis than any other of the valvular affections. A remarkable fact is the relative frequency of its occurrence in women, whether the basis of the estimate is post-mortem or clinical. Of fifty-three patients dying in St. Mary's Hospital and examined after death, thirty-eight were females and only fifteen males—seventy-two and twenty-eight per cent.

respectively. Of eighty-one cases collected by Hayden fifty-four were females and twenty-seven males—66·6 and 33·3 per cent. Sir Dyce Duckworth, in eighty cases, found no fewer than sixty-three women—78·75 per cent. It cannot be said that any satisfactory explanation of this great disproportion of women affected by mitral stenosis has been given. It is true that rheumatism is more common in girls than in boys, but were this the only reason, there ought only to be a general predominance of valvular disease in women, and not of this particular condition. Possibly the greater liability of girls to anæmia at the period of puberty may have some bearing on the greater incidence of mitral stenosis in the female sex, especially when it is borne in mind that anæmia is frequently attended with augmented arterial tension which, by increasing the stress on the mitral valves, is a cause of insidious damage. These two factors may tend to keep going the chronic inflammatory process, which results in adhesion of the cusps of the mitral valve, and in further constriction of its orifice by cicatricial contraction of the fibrous tissue thrown out around it. For mitral stenosis is a slowly progressive disease, and some time must elapse from the initial attack of endocarditis before an extreme degree of constriction is produced.

Before describing the physical signs it will be useful to briefly review the morbid anatomy and physiology of mitral obstruction.

As a rule the heart is not very large. Weights of ten and twelve ounces are common, and fourteen or fifteen will represent about the average. The left ventricle is usually not increased in size, the left auricle is dilated, and the right ventricle much enlarged and its walls greatly thickened. The left ventricle, however, may be enlarged and its walls thickened if mitral incompetence preceded the mitral stenosis, and there may therefore be great differences in the weight of the heart and dimensions of its cavities in different cases.

The Mitral Orifice.—The condition of the mitral orifice may vary greatly, and an extreme degree of stenosis does not seem to be incompatible with life. The orifice may be so constricted that it will scarcely admit a penholder, and very commonly it will only admit the tip of the little finger. To such conditions the term "button-hole" has been applied, but it is scarcely applicable, as the margins of the orifice are usually rigid and unyielding. In other cases the aperture may be funnel-shaped. Frequently there is no trace of the mitral valves remaining as such, as they are adherent at the margins, contracted down to form a rough irregular lining to what is left of the original mitral orifice.

The characteristic physiological effect of the lesion upon the heart is dilatation of the left auricle with more or less thickening of its wall, and great hypertrophy with some dilatation of the right ventricle. The left ventricle is not correspondingly enlarged, and may retain its normal size, while the right ventricle, by its growth, may displace it backwards, so that no part of it appears on the anterior aspect of the heart, and its apex is no longer in contact with the chest wall.

When the mitral orifice is narrowed, it is the left auricle and right ventricle only which are called upon to exert increased force, since there is no obvious cause of increased resistance in the systemic circulation. The same may be said when the valvular lesion is mitral incompetence with regurgitation; but here another element of change comes in, which makes a difference between obstruction and incompetence. The high pressure in the pulmonary veins and left auricle, which is a result of the damming back of the blood and of the increased force of the right ventricle, causes a forcible inrush into the left ventricle during diastole; and this, so long as the orifice remains of the natural size, must distend, and in the long run dilate its cavity, taking effect, as it does, during the unresisting period of the ventricular rhythm. But an increase in the capacity of the cavity multiplies by so much the force required to expel its contents, and this constitutes a demand for hypertrophy. We have, then, as a result of mitral incompetence, dilatation, and more or less hypertrophy of the left ventricle; but the hypertrophy here is required as compensation for the dilatation, and not to overcome any direct effect of the impairment of the valvular apparatus. In stenosis of the mitral orifice, the pressure which thus affects the left ventricle is intercepted by the narrowed orifice. There is scarcely time for it to be adequately filled during diastole, still less for any distending effect to be produced. We see, then, how it is that the left ventricle usually remains of normal size and does not increase *pari passu* with the right. Frequently, however, the left ventricle is dilated and more or less thickened, and here our reasoning appears to be at fault. But the difficulty disappears on reflection. Not uncommonly the change in the valves, which eventually glues them together and narrows the orifice, only interferes at first with their apposition permitting of regurgitation, which may, indeed, be for a long time the predominant result; and, in point of fact, incompetence often precedes the establishment of obstruction. We have here abundant cause for differences in the condition of the walls and cavities of the heart found after death, and, it must be added, for variations in the clinical history, and especially for diversity of physical signs.

Physical Signs. The Pulse.

The pulse of mitral stenosis is interesting. It is, according to my experience, almost always regular until the heart is obviously failing, unless the obstruction be complicated by regurgitation or by valvular affection of the right side of the heart.

The **artery** at the wrist is small, and is full between the beats, presenting the characters of moderately high tension—that is, it can be rolled under the finger and is not very easily compressed. In my experience this modified high tension pulse is almost constant, and it points to resistance in the capillaries; but the cause of such resistance is not readily perceived. It may be due to contraction of the arterioles, either reflex or by direct stimulation, consequent on the blood being charged with impurities due to imperfect elimination, or possibly owing to backward pressure in the veins, which makes itself felt through the capillary network. More probably, however, it is simply an effect of the contracting down of the entire arterial system owing to the diminished supply of blood from the imperfectly filled left ventricle.

When irregularity comes on, it is usually at first inequality in the force of the beats, without marked disturbance of the rhythm. Then some of the beats of the heart fail to reach the wrist—no doubt from inadequate filling of the ventricle; the action of the heart may thus continue to be regular when the pulse is irregular. In some rare instances, there is only one beat of the pulse for every two beats of the heart, the contraction of the left ventricle at every alternate systole being inadequate to raise the aortic valves; and, on listening to the heart, there will be no aortic second sound with the alternate systoles, the rhythm as expressed by the sounds being one—two—*one*, one—two—*one*, with an accent on the third sound heard. The heart-beats follow each other in couples, and the accentuated first sound, unaccompanied by an aortic second sound, is mainly produced by the abrupt contraction of the right ventricle.

The Heart.

The cardiac sounds and murmurs are varied, and sometimes perplexing; but it has appeared to me that they afford a means of estimating approximately the degree of constriction which the mitral orifice has undergone. The contraction does not take place all at once, but increases by slow degrees through many months or years, and it is to be expected that corresponding change in the physical signs will accompany this change of mechanical conditions. The physical signs are not the same in a given case from beginning to end, and by following the modifications of the sounds and murmurs which gradually supervene, I have been led to recognize three stages of the disease.

The heart is not usually greatly enlarged; the apex is displaced to the left and sometimes also downwards, but it is found, as a rule, not far from the normal situation.

The dilatation of the left auricle gives rise to an extension of dulness outwards, along the fourth and third left intercostal spaces, and the dilatation of the right auricle causes dulness up to or beyond the right border of the sternum.

The apex beat is not well defined, and in advanced cases a sharp and often powerful shock is felt on palpation, which, however, has not the deliberate thrusting character of hypertrophy, but is more like a tap. A thrill, presystolic in time, may usually be felt at the apex or just internal to it. The impulse of the right ventricle is powerful, lifting the lower end of the costal cartilages, and making itself seen and felt in the epigastrium.

Changes in the dimensions of the heart, however, have not the same direct relation to the degree of valvular mischief in mitral stenosis as in other valvular diseases, and it is by means of auscultatory signs that the division into stages is effected.

The pathognomonic sign of mitral stenosis is usually given as a presystolic murmur heard over a limited area at and to the inner side of the apex beat, between it and the left border of the sternum. It is not a smooth, blowing murmur, but has a rough, vibratory character, and is often accompanied by a thrill perceptible to the hand at the same spot. A further distinctive feature is the way in which it runs up to and suddenly ends in the first sound, which tends to become short and sharp, and is itself highly characteristic.

Sometimes in children after an attack of pericarditis, possibly as a result of pericardial adhesions, or in association with a mitral systolic murmur, a kind of rumbling presystolic murmur is audible, when no constriction of the mitral orifice exists. This murmur has not, however, the vibratory character of the true murmur of mitral stenosis; nor does it terminate abruptly in the first sound; nor is the first sound modified in the manner described above. Reduplication of the first sound should not be mistaken for a presystolic murmur.

The **pulmonic** second sound is accentuated, as a result of the backward pressure in the pulmonary circulation, and not unfrequently there is reduplication of the second sound, owing to want of synchronism in the closure of the pulmonic and aortic valves.

The characteristics of the three stages in the progress of the disease, as defined by auscultatory signs, will now be discussed.

In the **first stage** a second sound, as well as a presystolic murmur and first sound, will be audible at and to the left of the apex. It is the persistence of the second sound at the apex which is the chief distinctive feature of this stage.

Under these conditions I have never known serious symptoms to arise from the condition of the heart, and I have seen illnesses of different kinds, even serious attacks of bronchitis, passed through without the intervention of embarrassment of the circulation. It is very rarely that patients are admitted into hospital presenting simply the signs above enumerated, but they are frequently met with in out-patient practice and in consulting-rooms.

The Second Stage.—This is marked by the disappearance of the second sound at the apex, and by the short, sharp character of the first sound, which also usually becomes very loud; the first sound, in effect, comes to resemble a second sound. Mistakes in diagnosis may now be easily made. In mitral stenosis, at this stage, and in mitral incompetence there is alike heard at the apex a murmur followed by a short, sharp sound; but in the former the murmur is presystolic in time, and the sound is the modified first sound, while in the latter the murmur is systolic and the sound is the second sound. Very slight attention would, in most cases, suffice to prevent any confusion between the two, but an apex murmur is liable to be set down as the familiar systolic murmur of regurgitation without further investigation, and thus mitral stenosis, the most serious of the diseases of the valves, at a period, too, when symptoms may be impending, is taken for incompetence, which is attended with less danger than any other of the valvular affections. To bear this source of error in mind is to avoid it; but cases are sometimes met with in which, from absence of cardiac impulse and from the similarity between the sounds, it is not easy to follow the rhythm of the heart and time the murmurs and sounds. Flexible stethoscopes are here at a disadvantage as compared with the rigid wooden instrument, which communicates to the ear and head, not only sound, but a sense of shock which at once indicates the moment of the systole, and this when there is no impulse perceptible to the hand. To attempt to time the first sound by the radial pulse is, of course, fallacious. The carotid pulse is a safer guide, but it is not always easy to co-ordinate tactile and auditory impressions.

The most trustworthy method of determining the relation of sounds to the cardiac rhythm is to find a spot in the region of the base where the first and second sounds are unmistakably recognized, and then from this point to follow the sounds, step by step, toward the apex, when it will be found which of them it is that disappears, or which maintains some distinguishing peculiarity.

The presystolic murmur itself usually undergoes a modification in this stage. As commonly heard, it occupies the end of the diastolic interval, running up to the first sound, and corresponding therefore with the auricular systole. But it may be a long or a short murmur, and may indeed occupy the entire diastolic interval, when it will correspond, not only with the auricular systole, but also with the active diastolic rebound of the ventricle from contraction; that is, the murmur is produced by the current of blood sucked into the ventricle as it dilates, as well as by that driven into it by the systole of the auricle. This murmur, occupying the whole of the diastolic interval, is most frequent in bad cases. But it may further be cut in two, and, instead of being a continuous rumble from the second sound up to the first, may subside as the active dilatation of the ventricle ends, and before the auricular systole begins; or the proper presystolic part of the murmur may be lost while the diastolic part remains audible, so that the only murmur heard is diastolic, *i.e.* a diastolic mitral murmur, strictly speaking. These varieties of murmur may be represented diagrammatically as shown in Fig. 11.

The disappearance of the second sound at the left of the apex, which, with the short, sharp character of the first sound, marks the second stage of mitral stenosis, is probably explained by the following considerations. In the normal heart, a second sound is always audible at and to the left of the apex; and repeated careful examinations have convinced me that it is the aortic second sound which is here heard, and not the pulmonic, even when this is accentuated and unduly loud. In mitral stenosis, there are two influences tending to prevent the aortic second sound from reaching the surface of the chest. First, the left ventricle, not undergoing dilatation and hypertrophy, while the right enlarges greatly, is overlapped by the latter, which usurps the position of the apex, and thus the left ventricle cannot conduct the aortic sound to the chest wall. Again, the

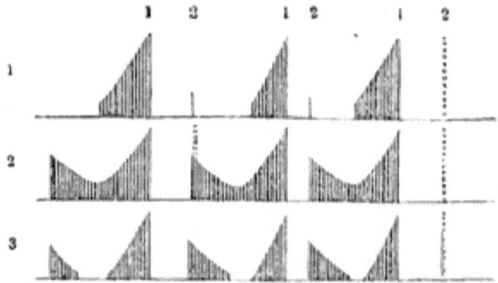

FIG. 11.—1. PRESYSTOLIC MURMUR CORRESPONDING WITH AURICULAR SYSTOLE. 2. MURMUR OCCUPYING WHOLE DIASTOLIC INTERVAL. 3. MURMUR DIVIDED INTO DIASTOLIC AND PRESYSTOLIC PORTIONS.

aortic second sound will be weak, because the diminished amount of blood entering the ventricle, in consequence of the narrowed mitral orifice, will not distend the aorta, and will, therefore, fail to produce a powerful recoil such as is necessary for the production of a loud second sound.

The modification of the first sound in mitral stenosis is remarkable, and in searching for an explanation of this, one is struck by the analogy of the sharp, sudden, and tapping character of the apex beat to the shortness and sharpness of the first sound, which would seem to imply that some common cause has been instrumental in the production of these peculiarities in both instances. If this be the case, it is clear that the ventricular wall must be one of the factors, and the following explanation may be suggested. Owing to the narrowing of the mitral orifice there is not time in the diastolic interval for a sufficient amount of blood to flow into the left ventricle to completely fill it. At the commencement of systole, therefore, the ventricular cavity is not fully distended with blood, so that the muscular walls at the first moment of their contraction meet with no resistance; then, closing down rapidly, they are suddenly brought up and made tense as they encounter the contained blood. It seems plausible that this sudden tension of the muscular walls and the abbreviated systole of the left ventricle, would account both for the sharp and tapping apex beat as well as for the short first sound.

In severe palpitation and tachycardia the first sound is also extremely short and sharp, and probably a similar explanation holds good here, namely, that the brief diastolic interval which is the rule in such cases does not afford sufficient time for the complete filling and distending of the ventricle with blood.

The **third stage** is characterized by the disappearance of the presystolic murmur, so that, the second sound being already lost, the only sound at and outside the apex is the short sharp first sound described. This is not unlike the first sound heard in dilatation with thinning of the left ventricle; but the absence of the second sound to the left of the apex constitutes a diagnostic difference, since this is distinct in dilatation.

No careful observer who has devoted much attention to the study of mitral stenosis has failed to notice that the presystolic murmur is sometimes absent in cases in which an advanced stage of this condition is met with after death. The principal justification, however, for taking the disappearance of the presystolic murmur as a distinct stage in the clinical history of mitral stenosis is that, very commonly when pulmonary complications set in, or other serious symptoms arise, the presystolic murmur is lost, and that it again becomes audible when these subside and the patient improves.

It is a matter of repeated and familiar experience for cases to be admitted into hospital on account of serious symptoms with only the short sharp first sound audible at the apex, and to leave after recovery with a presystolic murmur. The third stage, however, is not necessarily attended with serious symptoms, though this is the rule.

The probable cause of the disappearance of the murmur is the establishment of tricuspid incompetence. The giving way of the tricuspid valve and the occurrence of considerable reflux into the right auricle, make it impossible for the right ventricle to sustain the same high pressure in the pulmonary circulation and left auricle as was present previously. There is not, therefore, sufficient pressure to force the blood through the mitral orifice rapidly enough to generate a murmur.

Symptoms.

The patient, up to a very advanced stage, has not the look of heart disease, is neither pallid, nor dusky, nor anxious-looking, but often has a good bright colour and cheerful expression. There is usually some breathlessness on exertion, but not unfrequently, up to the moment when, from some cause or other, serious symptoms set in, he or she is unconscious of embarrassment of the circulation, and capable of ordinary work. On the other hand, such patients are sometimes liable to congestion of the lungs and hæmoptysis after exertion. Mitral stenosis is, moreover, the valvular affection which, independently of inflammation or ulceration, most frequently gives rise to arterial embolism. This is not the result of detachment of fragments from the valves, but is due to the formation of fibrinous coagula between the musculi pectinati of the left auricle, or in the appendix, or, less frequently, between the columnae carneae of the ventricle, which are shaken out of their beds by exertion or excitement, and carried into the circulation. Another condition more frequently met with as a consequence of mitral stenosis, according to my experience, than in other valvular affections, is great enlargement of the liver, with true pulsation of this organ; and it is not uncommon to find fluid in the peritoneal cavity before there is œdema of the feet and legs, or the œdema will disappear with rest in bed, while ascites remains for a time; whereas cardiac dropsy, as a rule, begins in the connective tissue of the most dependent parts.

Diagnosis.

In the first stage, where the vibratory presystolic murmur ending in the short sharp first sound, and the second sound are present, there will be little difficulty in arriving at a diagnosis. The short, rumbling, presystolic murmur often met with during or shortly after an attack of peri- or endo-carditis, more especially in children, may, however, lead to some confusion. This murmur, however, is not of the loud vibratory character of the typical presystolic murmur, but is usually short and rumbling; it does not indicate the presence of mitral stenosis, at any rate, in the sense of an established organic lesion. A further distinctive feature in diagnosis will be the absence of the typical modification of the first sound, which, instead of being short and sharp, will be dull, low-pitched, and possibly reduplicated.

In the second stage, when the second sound is lost at the apex and only the presystolic murmur and first sound are present, these may possibly be taken for a systolic murmur and second sound which the first sound has come to resemble in character; the case would then be mistaken for one of mitral incompetence in which the prognosis is much less serious. Due care and accurate timing of the sounds will prevent this. Confusion is, however, especially liable to occur when mitral regurgitation is present as a complication. It is then possible to mistake the presystolic and systolic murmurs together for a prolonged systolic murmur, the short sharp first sound being taken for an accentuated second sound. But attention to the character and time of the murmurs will obviate such a mistake. The systolic murmur of mitral regurgitation is blowing or musical, and begins with an accent, whereas the presystolic murmur is vibratory and ends with an accent. Accurate observation of the time at which the short sharp sound is heard will also prevent the mistake, but this will not perhaps be so easily distinguished as the character of the murmurs.

They may be represented diagrammatically as follows :—

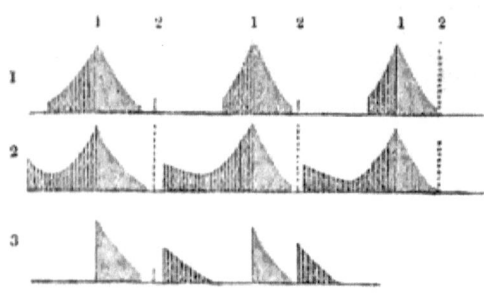

FIG. 12 (Dark shading = presystolic murmur; light shading = systolic murmur).
1. ORDINARY PRESYSTOLIC WITH SYSTOLIC MURMUR. 2. PRESYSTOLIC MURMUR OCCUPYING THE WHOLE OF DIASTOLE, WITH SYSTOLIC MURMUR. 3. DIASTOLIC PORTION ONLY OF PRESYSTOLIC MURMUR WITH SYSTOLIC MURMUR.

In the third stage, complicated by mitral regurgitation, recognition of mitral stenosis will often present great difficulties. There will be no presystolic murmur, and only a first sound with a systolic murmur will be audible at the apex. The absence of the second sound will be of great value in identifying the stenosis. If, however, the mitral regurgitation has preceded the onset of stenosis, it may have given rise to a degree of hypertrophy and dilatation of the left ventricle, such that the apex-beat, which comes into contact with the chest wall, is still that of the left ventricle and not of the right, so that the aortic second sound is still conducted to the chest wall. In such cases the modification of the first sound is an important aid to diagnosis; regurgitation tends to destroy the first sound, stenosis to shorten and intensify it, so that the presence of a short sharp first sound with a systolic murmur, in a case where there is obviously serious cardiac mischief, should at once suggest the probability of the presence of mitral stenosis.

When, however, compensation has broken down in a case of combined stenosis and regurgitation, extraordinary fluctuations in the physical signs may be present; at one time a systolic, at another a presystolic murmur, at another time both may be heard, or perhaps only a short sharp sound alternating with murmurs, which it is impossible to time. At the same time the action of the heart will be rapid and markedly irregular. In such cases the extraordinary fluctuations must themselves give the chief clue to the diagnosis, though it will be impossible to estimate the extent of either lesion.

It would appear that when the high pressure in the left auricle is well sustained it generates a presystolic murmur, and prevents regurgitation, so that no systolic murmur is generated, and there is no reflux, even though the typical "buttonhole" mitral orifice is present, in which closure of the orifice is absolutely impossible, as no valves in fact remain as such. When, on the other hand, the pressure in the left auricle and pulmonary circulation falls below a certain point, either from break-down of the right ventricle or from the onset of tricuspid regurgitation, mitral reflux takes place during systole, and there is not sufficient pressure in the left auricle during diastole to generate a presystolic murmur. We may thus explain these extreme and confusing varieties in the physical signs of the third stage of mitral obstruction complicated by regurgitation, and account for the absence of a systolic murmur in many cases where, from the condition of the mitral orifice post mortem, we should have expected mitral regurgitation to have been present during life.

In **aortic incompetence** a presystolic murmur is sometimes present in addition to the diastolic murmur, which does not necessarily imply the existence of mitral stenosis. The probable significance of this murmur and the explanation of its presence have already been discussed in the chapter on Aortic Incompetence.

Prognosis.

Mitral stenosis stands next to aortic regurgitation among valvular affections in the order of gravity. The average age at death, as deduced from 53 cases abstracted from the post-mortem records at St. Mary's Hospital, was found to be 33 for males and 37 or 38 for females, which is higher than one would expect.

A suggestive inquiry is, why mitral stenosis should be so serious, and especially why it should be attended with greater danger to life than mitral incompetence. One reason is, that the effects of obstruction here are not so easily neutralized as are the effects of regurgitation. The high pressure maintained in the pulmonary circulation and in the left auricle, by hypertrophy of the right ventricle, will antagonize incompetence of the valve in two ways— by resisting the reflux during systole, and by more rapid filling of the ventricle during diastole. In stenosis this high pressure can only be of service by increasing the rapidity of the passage of blood through the narrowed orifice during diastole; and, as diastole only lasts for a certain time, if the contraction be extreme, the ventricle cannot be properly filled before the systole is again due. When such is the case, the compensation is inadequate. On the other hand, since the constriction may reach a degree which, in the absence of experience, would seem quite incompatible with life, and since the subjects of such extensive change must have lived through all the intermediate degrees, it can scarcely be simply that obstruction, as such, is specially dangerous. Probably the explanation of the greater danger of obstruction, as compared with regurgitation, lies in the fact that, when once adhesion between the flaps of the valves has set in, it tends to go on, the friction and strain keeping up chronic inflammation, which gives rise to further adhesion of the valves and contraction of the orifice.

When mitral stenosis is established in childhood or early adolescence, the prognosis is more serious than when it occurs in later life. This is partly owing to the progressive tendency of the constriction of the orifice, which is more marked in early life, and partly to the fact that the stenosed orifice does not increase in size while the growth of the heart continues, so that even though no further actual narrowing takes place, the relative disproportions between the mitral orifice and the cavities of the heart will increase as the heart attains its full development.

The important point in prognosis, however, is the comparative prospect of life in individual cases, and it is in the estimation of this that the recognition of the different stages is of service. The extent of the hypertrophy and dilatation of the walls and cavities of the heart does not afford very great assistance, as we have seen that mitral stenosis does not give rise to any marked change in the left ventricle; a certain amount of information, however, as to the amount of obstruction may be gathered from the degree of hypertrophy and dilatation of the right ventricle, on which has fallen the work of overcoming the obstruction. A careful study of the sounds and murmurs will give more precise information, so that the three stages in the disease, as described above, must be borne in mind in attempting to form a prognosis. So long as the second sound is heard at and beyond the apex, there is little or no liability to the occurrence of symptoms, and there is no immediate danger. It must, however, be borne in mind that, owing to the tendency of the stenosis to increase, the prognosis as to prolonged life in the future is not favourable in the majority of cases.

When the second sound is lost at the apex, that is, when the second stage is reached, there may still be immunity from symptoms under ordinary conditions of life; but there is no capacity for the adjustment of the circulation to deviations from these, so that any imprudence or slight over-exertion, or even mental worry, on the part of the patient is liable to bring about a break-down of compensation. One must not, therefore, be thrown off one's guard by absence of complaints, or by the apparent good health of the patient. With suitable precautions and care, however, the condition of the patient may remain the same for years, though any tendency to increasing shortness of breath or any fresh attack of rheumatism will be reasons for apprehension, as indicating that the constriction is increasing.

When the third stage is reached and serious symptoms, such as dyspnœa, dropsy, pulmonary apoplexy, and other effects of venous congestion and over-distension of the right side of the heart, are present, the first element in the mental calculation will be the severity of the symptoms. But recoveries are witnessed in conditions apparently so desperate that if it is the first time the patient has suffered

from a similar complete break-down of compensation, the
case must not be pronounced absolutely hopeless. The
number of times severe symptoms have arisen and the
readiness with which they have been provoked become the
most important considerations. If the patient has had
similar attacks previously, and if a very slight cause has
been sufficient to induce them, the danger is very great,
and the chance of even temporary recovery is a poor one.

TREATMENT.

The first stage of mitral stenosis is attended with few
symptoms, and rarely calls for treatment. So long as the
narrowing of the orifice is only moderate, compensation
appears to be easily effected, and the patient suffers little
or no inconvenience. The fear is that once the flaps of
the valve have begun to adhere, the adhesion will extend
and gradually narrow the fissure between them. If this
could be prevented by any means, the treatment effecting
it would be of extreme value. We have no power, however, of directly influencing the process of adhesion, and
can only endeavour to obviate causes tending to keep up
or excite inflammation of the valves. Every possible precaution should be taken against rheumatism, which is
extremely prone to attack the damaged valves; and slight
rheumatism, which might be neglected in a sound individual, should receive attention in a patient suffering
from mitral stenosis of however small degree. Unduly
high pressure in the arterial circulation will react on the
valves by giving rise to high intra-cardiac pressure, and
stress upon them will tend to keep up or revive irritation.
Over-eating and drinking, again, and constipation, giving
rise to accumulation of impurities in the blood, will have
a like tendency. Precautions based on this knowledge
should therefore be inculcated upon the patient.

Bronchitis and other affections of the lungs will increase

the resistance in the pulmonary circulation and the strain on the right ventricle, and should be specially guarded against.

As the narrowing of the orifice progresses, or when the affection is found on examination to have reached the second stage, and when symptoms such as breathlessness and cardiac pain or oppression or weight are readily induced, or are more or less constantly present, or when hæmoptysis has occurred, the precautions suggested above should be urged more emphatically. Mercurial purgatives will then be of service, and rest in the recumbent position, and, if necessary, confinement to bed for a time should be ordered. Strychnia and iron, with nitroglycerine or nitrites and stimulants, may be prescribed, but digitalis should not be given unless there are symptoms of right ventricle failure, and not then until after free purgation; on no account should it be given for a long period.

The rules for exercise will be the same as in other forms of valvular disease. Sufferers from mitral stenosis of slight or moderate degree are perhaps more liable to do themselves harm by imprudent exertion than the subjects of mitral incompetence, as they are often not checked by breathlessness, but persist in overtaxing their strength till severe pain in the heart, or perhaps an attack of hæmoptysis is induced. At an advanced stage of the disease exertion may prove suddenly fatal, but never while the patient is free from serious symptoms; more commonly it is by overthrowing the compensatory balance already inclined to the wrong side, and so aggravating the existing venous stasis, or by detaching a thrombus, which gives rise to embolism, that effort or excitement proves injurious and ultimately fatal. Hæmoptysis is rarely considerable, and seldom requires any other treatment than rest and aperients.

When from neglect of precautions, or from overtaxing of strength by unavoidable duties, or from depressing emotions, or from the advance of the disease, decided symptoms of right-ventricle failure have supervened, such as great weakness, cough, dyspnœa, with evidences of venous stasis in swollen jugulars and enlarged liver, and dropsy, then energetic measures will be required. These will be mainly such as will relieve the venous engorgement and the overloaded right ventricle and auricle, smart purgation by calomel or blue pill and colocynth, followed, if necessary, by salines.

When the symptoms are very urgent, venesection may be of striking service, and in certain cases there is no other treatment that will take its place and avail to avert a speedy fatal termination.

The indications will be, not a full bounding pulse, but the opposite—a small, weak, irregular pulse, many of the beats being scarcely perceptible; the heart, on the other hand, more especially the right ventricle, will be beating violently, epigastric pulsation being very marked, while the apex beat is scarcely perceptible. The liver will be enlarged, and perhaps pulsating, and the jugulars will be full and pulsating. The contrast between the powerful right ventricle impulse and the small, weak, irregular pulse is very striking, and is one of the most important indications for venesection. The sufferer will be in a state of dyspnœa, though not always of an agonizing kind, and not always compelling him to sit up; the face may be dusky and the lips blue, but it may also be pale with a red patch of injected capillaries on the cheeks.

The narrowed mitral orifice constitutes a fixed obstacle, which keeps up an unremitting backward pressure in the pulmonary circulation, and makes it difficult or impossible for the right ventricle to overcome the paralyzing overdistension to which it is subjected. Venesection lessens the amount of blood arriving by the veins, and gives the right ventricle a chance of recovery, so that it can again contract down more or less efficiently on its contents.

Venesection does not, however, dispense with the necessity for relieving the portal circulation by purgation. Stimulants which without these measures afford no relief, and may indeed do harm, will then be of the greatest service, and digitalis and like remedies will find their opportunity.

When the symptoms are less urgent or venesection is objected to, seven or eight leeches applied over the liver may be of service, and in hospital patients, where rest and care and proper nourishment make such an enormous difference in the influences acting on the patient, leeches will usually be sufficient.

The administration of digitalis in the early stages of the disease is seldom if ever called for; it is only when there are symptoms of right ventricle failure, and then only after free purgation and, if necessary, venesection have been employed, that it should be prescribed. Up to a certain point in such cases its influence is often most beneficial, but sometimes it fails to relieve, and even appears to aggravate the symptoms. If continued too long in cases where it has been of signal service, unfavourable effects may supervene, marked by slowing of the pulse, a sense of præcordial oppression, and by coupled heart-beats, the first of which alone reaches the wrist, the second being unaccompanied by an aortic second sound.

Digitalis, therefore, must be employed with caution in mitral stenosis, and its effects should be carefully watched. Under no circumstances should it be prescribed unless the patient is under observation, and it should rarely be given for a long period of time.

Nitro-glycerine and other vasodilators may sometimes be given with good effect for many weeks or even months in conjunction with general tonics, such as iron, quinine, and nux vomica.

CHAPTER XIV.

VALVULAR DISEASE OF THE RIGHT SIDE OF THE HEART.

TRICUSPID INCOMPETENCE AND STENOSIS—PULMONIC INCOMPETENCE AND STENOSIS—SYSTOLIC PULMONIC MURMURS WHICH DO NOT INDICATE STENOSIS.

PRIMARY valvular disease is rare, and for the most part is congenital.

Tricuspid regurgitation is so common when the right ventricle is overdistended by violent exertion (the so-called safety-valve action) that it may be looked on as physiological; it is not usually attended with a murmur. Tricuspid incompetence, again, with or without a murmur, is an early and almost constant effect of back pressure through the lungs when there is serious valvular disease of the left ventricle. In both instances the cause of the regurgitation is dilatation of the right ventricle, temporary or permanent, as the case may be.

The tricuspid valve may, however, be damaged during intra-uterine life, or more rarely in childhood or adolescence by endocarditis of rheumatic origin, or, where serious valvular disease of the left ventricle exists, may undergo thickening and contraction from chronic inflammation set up by the irritation and undue strain caused by protracted high pressure in the pulmonary circulation.

The **murmur** attending tricuspid regurgitation is systolic in time and is usually blowing in character, having its maximum intensity about one-third of the distance between the left edge of the sternum and the vertical nipple line. It is usually audible outwards towards the apex, and sometimes at the apex itself, where it may be mistaken for a mitral murmur. Such a murmur, when constant and not occasional only, may be looked on as indicative of definite tricuspid insufficiency, with probably actual change in the valvular flaps and chordæ tendineæ. A musical tricuspid systolic murmur may sometimes be heard over a limited area, but it seldom has any important significance.

It is not, however, from the character of the murmur, or from its presence, that conclusions as to the degree of tricuspid incompetence are to be drawn. More important information is gained from the condition of the veins of the neck and from enlargement of the liver. The veins of the neck are more or less distended according to the degree of regurgitation, and the external jugulars may attain even the size of the little finger. Frequently they will fill from below when emptied by pressure.

Pulsation is usually present when there is much regurgitation, and is often very conspicuous. It is sometimes seen to be double, the contraction, first of the auricle, then of the ventricle, sending a reflux wave along the jugulars. At times the pulsation in the internal jugular vein is so marked, and extends so high, that at first sight it may be taken for the carotid throb of aortic regurgitation; but it is of course easily extinguished by light pressure.

The effect of tricuspid incompetence, causing damming back of the blood in the inferior vena cava, is felt by the liver, which gradually becomes much congested and enlarged, and eventually may pulsate as the reflux becomes more considerable, and the right ventricle begins to fail.

TRICUSPID STENOSIS.

Tricuspid stenosis is usually associated with mitral stenosis, and may be looked on as secondary to it in some cases, due to chronic inflammatory changes set up by back pressure through the lungs and strain on the tricuspid valves. It is, however, worthy of notice that tricuspid stenosis is rarely associated with pure, uncomplicated mitral incompetence, although there may be an equal degree of back pressure through the lungs. Tricuspid stenosis may also be present at so early a stage in association with mitral stenosis, that it could not be regarded as a secondary affection. In such cases the tricuspid valve must have been damaged by the same attack of endocarditis which injured the mitral valve. Tricuspid stenosis is rarely if ever found as an isolated valvular lesion.

The physical sign, when it can be recognized, is a presystolic murmur, audible in the tricuspid area, but it is not easily distinguished from a concurrent mitral presystolic murmur. In several cases in which tricuspid stenosis has been diagnosed from the symptoms during life and found on post-mortem examination, and when consequently the presystolic murmur has been carefully and perseveringly sought for over a long period, it has been impossible to recognize and distinguish it. There can be no doubt, however, that the murmur has been frequently heard and recognized. Another important physical sign of tricuspid obstruction is distension of the jugular veins with little or no pulsation. Dr. Mackenzie of Burnley has obtained graphic records of jugular pulsation, and describes an auricular and ventricular type: when the right side of the heart is distended, the double jugular pulsation—the first wave auricular and the second ventricular—is frequently very distinct. But if the tricuspid valves are contracted and rigid and the orifice narrowed, the ventricular wave is cut off, and the auricle being at the same time very often paralysed by over-distension, the auricular wave may also be missing.

When any considerable degree of tricuspid stenosis exists, the symptoms of embarrassment of the circulation and of venous back pressure, are usually present in an extreme degree, and dropsy is the rule. In uncomplicated mitral stenosis dropsy is rarely present till the final stage of the disease, when there is complete cardiac failure. When, therefore, dropsy supervenes at an early period, we are justified in suspecting the presence of tricuspid stenosis in addition to mitral stenosis, even though there may be no murmur or other definite physical sign to point to it.

The Pulmonic Valves.

Disease of the pulmonic semilunar valves is very rare, and of the two conditions, insufficiency and stenosis, the former is the more uncommon.

Pulmonic insufficiency gives rise to a diastolic murmur, best heard in the left third intercostal space, and conducted downwards. It must, however, be borne in mind that the murmur of aortic regurgitation is also frequently heard at this spot, so that before venturing on a diagnosis of pulmonic regurgitation, it must be ascertained, not only that the pulmonic second sound is impaired, but also that the carotid throb and collapsing pulse are absent, and that the aortic second sound is unimpaired. No special train of symptoms can be attributed to pulmonic regurgitation.

Pulmonic stenosis is nearly always a congenital defect, and as it is fully discussed in the chapter on Congenital Malformations, little need be said here. The murmur to which it gives rise is systolic in time and usually loud and rough, varying in intensity. It is most distinct in the third left space, about three-quarters of an inch from the margin of the sternum, but it is conducted along the branches of the pulmonary artery and is often audible over the whole cardiac area and far beyond, sometimes over the entire chest front and back. Pulmonic stenosis is often associated with some other congenital defect, such as a perforate interventricular septum or patent foramen ovale. When the last-named condition exists cyanosis is, in the majority of cases, present in a more or less pronounced degree, or is easily induced by exertion.

Systolic pulmonic murmurs are very common without change in the orifice or valves.

A pulmonic murmur may be present in a patient suffering from anæmia and disappear as the blood regains its normal character; it is then properly termed "hæmic."

Again, a murmur in this situation may be induced by severe or protracted exertion, and last for some hours or days. There will probably be at the same time indications of dilatation of the right ventricle, and though it is not easy to explain how this could give rise to the pulmonic murmur, we must accept it as the cause.

A third variety of systolic murmur in the pulmonic area, which may be loud and rough, is not unfrequently met with in adolescents of both sexes, which cannot be attributed to anæmia or to dilatation of the right ventricle. It is not indicative of temporary weakness or organic unsoundness of heart, nor is it incompatible with capacity for vigorous and sustained exertion, as boys and men in whom it is present can play football and train for races with impunity. Its presence may be made a pretext for rejecting candidates for the army on medical grounds, but I have known many men in whom such a murmur was present go through arduous campaigns without breaking down. I have never known such a murmur develop into actual heart disease; on the contrary, it usually disappears in adult life, except now and then in women.

The causation of such a murmur is apparently as

follows. Usually the conus arteriosus of the pulmonary artery is covered by the thin edge of the overlapping left lung. In the cases under consideration the covering by lung is incomplete, and a part of the conus comes into contact with the chest wall, and during systole is flattened more or less against the chest wall. An eddy is thus formed in the current of blood rushing into the pulmonary artery, which gives rise to a murmur. Evidence in support of this explanation is afforded by the fact that the murmur usually disappears when the patient is told to take a deep breath and hold it, as a cushion of lung is then brought over the conus arteriosus between it and the chest wall. The result is the more striking if the murmur happens to be loud and vibratory.

CHAPTER XV.

CONGENITAL MALFORMATIONS.

VARIETIES OF CONGENITAL MALFORMATIONS—RELATIVE FREQUENCY OF OCCURRENCE—OF SINGLE AND COMBINED DEFECTS—CYANOSIS—CAUSE OF CYANOSIS—PHYSICAL SIGNS—SYMPTOMS—DIAGNOSIS—PROGNOSIS.

SOME of the most important varieties of congenital malformations of the heart and great vessels are—

1. The heart consisting of two or three cavities, the interventricular or interauricular septum, or both, being absent. This is of very rare occurrence.

2. Incomplete interventricular septum, usually taking the form of a perforation in the upper third of the septum.

3. Patent foramen ovale.

4. Persistence of patent ductus arteriosus.

5. Stenosis of the pulmonary orifice due to constriction of the trunk of the vessel itself, or of the infundibular portion of the right ventricle, or to the adhesion or malformation of the valves.

6. Transposition of pulmonary artery and aorta. This is rarely found.

7. Malformations of the aortic, tricuspid, or mitral valves are not of common occurrence. In most cases, where the tricuspid valve is found to be affected at birth, the probabilities will be that it has been damaged by endocarditis occurring during fœtal life.

Of these varieties by far the most common is stenosis

of the pulmonary orifice. Of 181 cases of congenital malformation collected by Peacock, in 90 more or less contraction of the pulmonary orifice was present, and in 29 others the orifice or trunk of the vessel was obliterated.*

Deficiency of the interventricular septum usually takes the form of a perforation in the upper third of the septum, in the undefended or membranous space, so called because normally the septum here only consists of two layers of endocardium. It is rare as an isolated lesion, but is not uncommon in association with other malformations which give rise to unequal pressure in the two ventricles. For instance, it is frequently found coexisting with pulmonic stenosis, in which the pressure in the right ventricle is in excess of that in the left. Not unfrequently in such cases the right ventricle is greatly hypertrophied and the septum is deviated to the left, and in some cases the aorta arises wholly or in part from the right ventricle.

Patency of the foramen ovale may occur in connection with pulmonic stenosis; but this cannot be always attributed to excess of pressure in the right auricle, as it is sometimes found as an isolated lesion; and also cases are recorded in which the foramen was found closed, though excess of pressure must have existed in the right auricle.

The ductus arteriosus may remain patent in consequence of some obstruction to the passage of blood through the lungs or systemic vessels; not infrequently a patent ductus arteriosus occurs in association with a patent foramen ovale, the patency of both being due to similar causes; hence the two may be found together with, and as a direct result of, pulmonic stenosis.

In many instances of congenital malformation of the heart, the most marked and striking feature is the cyanosis of the patient; hence the various forms of congenital heart

* Peacock, "Malformations of the Human Heart," p. 193.

disease have been grouped together under the names morbus cœruleus, or blue disease, by English, and cyanose, or maladie bleue, by French authors.

The explanation of the cause of this peculiar discoloration is still a matter of dispute. Sénac, Corvisart, Gintrac, and others attribute it to the mixture of arterial and venous blood in the heart or great vessels, owing to defective septa or patent ductus arteriosus. Cruveilhier attributed it to venous congestion. It has been proved, however, that cyanosis may exist without the intermixture of currents of blood, also that complete intermixture may take place without the occurrence of cyanosis.

It is obvious, therefore, that neither of these explanations are sufficient. Stillé, with a view to determining the cause of cyanosis, collected 77 cases of congenital morbus cordis in which cyanosis was present. In 53 instances the pulmonary artery was constricted or impervious, or its orifice was obstructed in some way. He came to the conclusion, therefore, that cyanosis was due to venous congestion usually dependent on obstruction at the orifice of the pulmonary artery, or on some other cause giving rise to obstruction to the venous return. This is clearly not a complete explanation, as there is often no cyanosis in cases of morbus cordis in adults where the venous congestion is extreme. Peacock makes another important suggestion, that deficient aeration of the blood is a contributory cause of cyanosis; he says, "where only a small proportion of the blood is submitted at one time to aeration in the lungs, the whole mass must be of a dark colour, consequently the hue of the surface will be proportionately dark." This is especially the case in pulmonic stenosis, and Stillé's statistics, which show that pulmonic stenosis is the commonest cause of cyanosis, afford strong support to this view, that aeration of a small proportion only of the blood is the essential cause of cyanosis.

Physical Signs.

In **pulmonic stenosis** a loud, rough, systolic murmur is usually to be heard over the præcordial region with its maximum intensity at the level of the nipple, midway between the nipple and the sternum. It will be conducted along the branches of the pulmonary artery so that it will be heard over a large area on both sides of the chest; but it will be heard more distinctly on the left side of the chest between the base of the heart and the clavicle than over the aorta. In association with this the right ventricle will usually be hypertrophied.

Deficiency of the interventricular septum may give rise to a systolic murmur whose seat of maximum intensity is, according to Roger and Potain, in the fourth left space half an inch above the nipple; but as this defect commonly coexists with pulmonic stenosis, such a murmur would probably be masked by that due to the pulmonic lesion. A murmur that may occur in cases of deficient interventricular septum, and which I have met with sometimes, is one quite different in character to any that occur in the usual forms of valvular disease. It is harsh and loud, but its great peculiarity is that it never ceases, becoming suddenly louder and higher pitched with the systole, and subsiding into a continuous rumble during diastole, reminding one of the kind of noise of varying intensity made by a knife-grinder's wheel when a knife is being sharpened.

Patent foramen ovale.—There is no known auscultatory or other physical sign by which a patent foramen ovale can be diagnosed. Peacock quotes one case of patent foramen ovale without other defect, in which the patient, a girl, lived to the age of sixteen, when she died of pulmonary tuberculosis. There was no cyanosis, and till she began to suffer from tuberculosis there were no symptoms other than those of general debility. In a case under my care the patient was a man, who died at the age of thirty from bronchitis. There was no marked cyanosis during infancy, there were no special symptoms of morbus cordis, and no cardiac murmurs; the patient was, however, dull and of sluggish intellect, but could take long walks without seeming distressed in any way, or appearing the worse for it. During the fatal attack of bronchitis cyanosis developed, attended with torpor. A noteworthy point was that the cyanosis deepened during sleep, and there was no spontaneous waking up, and the patient was roused with difficulty. Eventually the torpor and cyanosis deepened into fatal coma. At the autopsy the only cardiac lesion found was a patent foramen ovale.

SYMPTOMS.

The child suffering from a serious congenital affection of the heart is usually irritable and fretful, and may be subject to fits. The fingers and toes are clubbed, and the extremities cold. He may be always cyanosed to a varying degree, or only become cyanosed on exertion. There will usually be constant shortness of breath, and paroxysms of dyspnœa may occur, in which cyanosis becomes so intense that the extremities become almost black. He will remain stunted in growth and backward in development, intellectually as well as physically. The symptoms will of course vary, according to the nature of the lesion. They will be most marked in a case of severe pulmonary stenosis, and may, as has already been seen, be entirely absent in a case of uncomplicated patent foramen ovale.

DIAGNOSIS.

Though in severe cases it is usually easy to arrive at a diagnosis of congenital heart disease from a history of cyanosis and dyspnœa since birth, with the clubbing of the fingers and toes, it is difficult and in many instances impossible to be certain of the exact nature of the malformation.

Where with a history of cyanosis and paroxysms of dyspnœa since birth we find a loud, rough, systolic murmur, with its maximum intensity on the left side at the level of the nipple, midway between the nipple and sternum, together with a hypertrophied right ventricle, we may be fairly certain that pulmonary stenosis is the main lesion; but whether a patent foramen ovale or perforate interventricular septum, or patent ductus arteriosus, is present as well, it will often be impossible to decide.

Prognosis.

The lesion is in this case stationary, and not progressive, but the malformations being widely different, the effect they have on the duration of life will vary considerably.

1. In **cases of moderate** constriction of the pulmonary artery without other malformation, for which hypertrophy of the right ventricle is sufficient to compensate, there will be no cyanosis, and the patient may live many years without serious inconvenience, except on violent exertion.

2. In cases where the foramen ovale is open the pulmonary stenosis is usually greater, and hence the duration of life will probably be less; but out of 20 cases which Peacock collected, 11 lived to the age of 15 years and over, 1 living to the age of 57.

3. Where with pulmonary stenosis the interventricular septum is also deficient, the prognosis is much less favourable, for not only must the pulmonary constriction be considerable to have given rise to this deficiency, but, further, in such case the aorta usually arises in part from the right ventricle. Of 64 such cases collected by Peacock, only 14 survived the age of 15. If, however, the degree of pulmonic stenosis is slight and the perforation in the

septum small, life may be prolonged, and two patients are now living, at the age of 30 or upwards, in which I believe this condition to exist; one has borne a child.

4. Where the pulmonary artery is impervious, the duration of life rarely exceeds a few months, though of 28 such cases of Peacock's, 3 lived to the age of 9 or 10, and 1 to 12 years.

5. Transposition of the main arteries, or arrest of development, so that the heart consists of two or three cavities only, is usually incompatible with life for any long period after birth, but 4 cases are recorded by Peacock where with one ventricle only and two auricles persons lived to the ages of 11, 16, 23, and 24 years respectively.

6. Patency of the foramen ovale uncomplicated by any other defect may not of itself give rise to any serious symptoms; doubtless in most instances where the opening is valve-like it will become closed before adolescence is reached. Where it persists or allows of extensive leakage, any lung affection, such as bronchitis, which tends to increase the pressure in the right side of the heart and cause a flow of non-aerated blood from the right to the left auricle, will be especially dangerous and liable to prove fatal.

CHAPTER XVI.

ADHERENT PERICARDIUM.

MORBID ANATOMY—PHYSICAL SIGNS—SYMPTOMS—DIAGNOSIS—PROGNOSIS—TREATMENT.

By the term "adherent pericardium" is implied the existence of adhesions between the visceral and parietal layers of the pericardium, the result of pericarditis. They may be limited to fibrous bands stretching across the pericardial cavity, or they may be universal, in which case the pericardium and heart are so intimately connected that the pericardial cavity is entirely obliterated. Adhesions may also exist between the chest-wall or pleura and the pericardium, as a result of so-called mediastino-pericarditis. The adhesions if of old-standing are tough and fibrous, so that the pericardium cannot be stripped from the heart without tearing the heart-substance. In the case of recent adhesions or lymph undergoing organization into fibrous tissue, the two layers of pericardium on being separated will present a honeycomb or bread-and-butter-like appearance, owing to the layer of thick, sticky lymph which coats the surface.

Physical Signs.

The physical signs differ according as the adhesions exist only between the two layers of the pericardium, or between the pericardium and chest-wall, or adjoining pleura as well. In the latter case they are more numerous and distinctive. Among them are the following:—

Fixation of the apex beat, so that it does not alter its position in deep inspiration and expiration or in change of posture of the body.

Systolic depression of one or more intercostal spaces to the left of the sternum, or of the lower end of the sternum and the adjoining costal cartilages, which may be caused by the heart dragging on them at each systole, through the agency of the pericardial adhesions. The systolic recession of spaces alone is, however, not a trustworthy indication, as it may be due to atmospheric pressure, more especially when the heart is much hypertrophied. When the costal cartilages or lower end of the sternum are dragged in there can be little doubt as to the diagnosis, as this could not be effected by atmospheric pressure.

Systolic recession of the site of the apex beat is an important sign when a definite apex beat can be felt; when there is no palpable apex beat, systolic pitting over its site may be due to atmospheric pressure.

A diastolic shock may sometimes be felt on palpation with the flat of the hand over areas on the chest-wall where systolic recession is present. It is due to the elastic recoil of the chest-wall at the commencement of diastole as soon as the pulling force exerted during the systole ceases.

Systolic retraction of the lower portions of the posterior or lateral walls of the thorax may indicate the presence of a universally adherent pericardium. Such retraction may, however, be seen though the pericardium is not adherent to the heart, but only to a larger extent than normal of the central tendon of the diaphragm and the muscular substance on either side, and to the chest-wall as well. In such cases the heart is usually greatly enlarged and hypertrophied from old valvular disease. The explanation seems to be that the portion of the diaphragm to which the pericardium is adherent is dragged upwards at each systole of the heart, so that the points of attachment of the digitations of the diaphragm to the lower ribs and costal cartilages are dragged inwards and retracted.

The **descent of the diaphragm in inspiration** may be interfered with by pericardial adhesions between the heart and diaphragm, more especially if the pericardium is adherent to the chest-wall in front as well. This will be shown by impaired movement in respiration of the upper part of the abdominal wall in the epigastrium and left subcostal region.

The **area of cardiac dulness** will be increased, and will remain unchanged in inspiration and expiration, where there are extensive adhesions between the pericardium and chest-wall, as the lung, which normally overlaps part of the heart, will have been pushed aside, or perhaps have become involved in the adhesions, and be collapsed.

Enlargement of Heart.—It is common with adherent pericardium to find the heart, more especially the right ventricle, considerably enlarged, in the absence of valvular disease or other obvious cause to account for it.

It seems probable that such enlargement may be indirectly due to pericardial adhesions as follows: The heart becomes dilated during an attack of pericarditis, and, before it recovers its tone or can contract down again to its normal size, the pericardium becomes adherent and fixes it in this condition of dilatation, the right ventricle suffering more than the left, owing to its thinner walls, as well as for other reasons.

Hypertrophy and dilatation of the heart, more especially of the right ventricle, may therefore, in the absence of other obvious causes, such as valvular disease, high arterial tension, etc., to explain it, be a physical sign of considerable importance.

Diastolic collapse of cervical veins was held by Friedreich to be of great diagnostic value when accompanied by systolic retraction of spaces; but I have never found it to be of service.

Systolic emptying of veins on the surface of the thorax may sometimes be observed, due to suction action, induced by the walls of the internal mammary veins being dragged apart by pericardial adhesions during systole of the heart.

When there are no adhesions between the pericardium and chest-wall the physical signs that may be present will be limited in number. There will be no recession of spaces except as the result of atmospheric pressure, no fixation of apex beat, no diastolic shock. As the pericardium is normally attached by fibrous bands to the central tendon of the diaphragm and to the muscular substance on either side of it, there may be some interference with the movements of the diaphragm in respiration. There may also be cardiac enlargement indirectly due to the adhesions, but in such cases a diagnosis will usually have to be made from other indications than physical signs alone.

Symptoms.

The symptoms in themselves are not in any sense characteristic. They are usually such as arise from cardiac embarrassment, more especially from the giving way of the right ventricle, such as œdema of the extremities, enlargement of liver, ascites, dyspnœa, etc.

Diagnosis.

The physical signs or symptoms of adherent pericardium, few of which may be present, are often in themselves insufficient to allow of a diagnosis being made, or even to arouse suspicion of its presence; but valuable help may be derived from careful consideration of the physical signs and symptoms together, and by balancing the former against the latter, so that the question is raised, "Do the physical signs present afford evidence of sufficient disease to account for the symptoms that have arisen?" When the symptoms are those of right ventricle failure, and are more severe than the physical signs present would lead one to expect, and have not been induced by undue exertion or imprudence, adherent pericardium must be thought of as being possibly responsible. For it is the right side of the heart more especially that is seriously hampered by pericardial adhesions, so that their presence may account for the unexpected breakdown of the right ventricle when the physical signs indicate that the valvular lesion is slight.

When with symptoms of right ventricle failure there is an absence of cyanosis, or of pulmonary congestion or lung mischief, this is further evidence in favour of adherent pericardium as a possible cause of the breakdown of the right ventricle. When by these means a suspicion of the presence of adherent pericardium has been aroused, confirmatory physical signs should be carefully sought for.

The above remarks apply to the question of diagnosis in cases where, with or without valvular disease, there is no history of pericarditis, and the adhesions are of old standing.

In cases of pericarditis, which can be kept under observation after the attack, there will be less difficulty in arriving at a diagnosis, and the indications which would lead one to suspect that the pericardium was becoming adherent are as follows:—

1. Prolongation of the attack of pericarditis evidenced by a harsh friction rub over the præcordial area, which may persist for some weeks. When at the margins of the area of cardiac dulness a pleuro-pericardial friction is also heard, it will indicate that adhesions are probably taking place between the pericardium and adjoining pleura or chest-wall as well.

2. Permanent enlargement of the area of cardiac dulness to a marked extent after the subsidence of the pericarditis.

3. The occurrence of symptoms of right ventricle failure after a period of temporary improvement, there being no

apparent exciting cause for the breakdown of the right ventricle.

Prognosis.

When the heart remains normal in size, and there are no adhesions between the pericardium and chest wall, the universal adherence of the pericardium to the heart may not in an adult tend to materially shorten life. When the heart is enlarged, or when the pericardium is also adherent to the chest wall, the prognosis is more serious. When adherent pericardium exists as a complication of valvular disease, it is still more likely to prove fatal eventually, by so hampering the right ventricle as to prevent its recovery when once compensation has broken down.

Treatment.

The discovery of adherent pericardium, when present, is important from the point of view of treatment, not because anything can be done to remedy or remove the pericardial adhesions, once they are formed, but because, when it is present, it will be necessary to impose additional restrictions on the patient, so that no undue risks may be run of upsetting the compensatory balance, which would only be restored with great difficulty.

CHAPTER XVII.

STRUCTURAL DISEASE OF THE HEART.

HYPERTROPHY—CAUSES OF HYPERTROPHY OF THE LEFT VENTRICLE—PHYSICAL SIGNS—SYMPTOMS—PROGNOSIS—TREATMENT—DILATATION—CAUSES OF DILATATION—ILLUSTRATIVE CASES—PHYSICAL SIGNS—SYMPTOMS—PROGNOSIS—TREATMENT.

THE muscular walls of the heart are liable to changes of various kinds, some of which constitute diseases which shorten life and give rise to much suffering. Of these structural alterations, some are extremely common—hypertrophy, dilatation, fatty degeneration; others rare—cancer, syphilitic gumma, abscess, aneurysm, localized fibrotic induration. We shall concern ourselves only with those which are comparatively frequent; the others, obscure as well as uncommon, are very seldom recognized during life, and a diagnosis is only made when an exceptionally clear case comes under the notice of an exceptionally acute observer.

Even when the common and familiar affections—hypertrophy, dilatation, and degeneration—only are taken into consideration, we find ourselves on much less secure ground than when dealing with lesions of the valves. The latter we can localize with great confidence; and knowing, partly by experience, partly by the application of mechanical principles, their effects and tendencies, we can, by making out how far such effects are manifest, form an opinion as to the probable course of the symptoms and as to the future of the patient. On examination after death, again, we can understand the connection between the lesion and the symptoms, and can follow the sequence of secondary changes in the heart and vessels which are set up by the original valvular defect. In the case of structural changes, on the contrary, the diagnosis cannot be made with the same precision, and we are often left in some degree in the dark even by a post-mortem examination. In one patient fatty degeneration has apparently proved fatal at so early a stage that the naked-eye characters of the condition are scarcely perceptible, and it is only by the microscope that its existence is definitely established; in another the change has proceeded so far that the fingers sink into the pale greasy walls, and the muscular fibres have almost disappeared, so that it is scarcely conceivable how the heart has been able to impress any movement whatever on the blood, or how life has been sustained through the intermediate stages of disintegration. So with regard to dilatation, there is no fixed relation between the degree of enlargement of the cavities and thinning of the walls found after death and the interference with the circulation observed during life. One man will live for years with a heart which has reached the extreme limits of dilatation, while another succumbs when it is but moderately advanced. If, therefore, we could make out with great exactness the dimensions of the heart, the size of its separate chambers, and the thickness of their respective walls, which is no easy task, we could not on these grounds alone compare one case with another, and decide upon the relative danger.

Many other considerations of extreme importance will come into the estimate—the functional vigour of the muscular walls as well as their thickness, the liability to palpitation, the state of the great and small vessels, the degree of peripheral resistance, the presence or absence of reflex irritation of the heart from gastric or other derangement. The question of prognosis thus becomes extremely complicated, and is beset with uncertainty. An element of chance or luck even comes in—the subject of advanced disease is at the mercy of the slightest accident; many a patient lives for years with a dilated or fatty heart who would be killed by an attack of influenza, or by a powerful emotion, or by tripping over a stone or a mat. A serious obstacle to the attainment of the minute and definite knowledge which alone is of real use in arriving at a sure prognosis in structural disease of the heart is the fact that a very large proportion of the cases occur in private and consulting practice, and comparatively few come under observation in hospital. It is difficult, therefore, to keep such cases in sight and to watch their progress, and seldom possible to obtain a post-mortem examination to confirm the diagnosis.

While, however, the great difficulty of prognosis and diagnosis in structural diseases of the heart is acknowledged, it is of the utmost consequence that such approximation to a forecast of the prospects of life as is possible should be an object of serious endeavour. The cases are numerous, far more numerous at and after middle age than those of valvular disease, and everything which has been said as to the importance of prognosis in general applies here. An early recognition of these changes, indeed, is often of greater service to the subject of them than in the case of valvular affections, since it reveals also the tendencies which are in operation, and often at a time when they can be successfully combated by treatment.

Too commonly, however, no attempt is made to recognize the existence and extent of degeneration or dilatation. The symptoms due to derangement of the circulation force themselves upon the attention of the medical man, but no murmur being detected, the only diagnosis ventured upon is that of "weak heart," a vague term which covers the entire ground, from temporary functional debility to disease inevitably and imminently fatal. Such a diagnosis reacts unfavourably upon the mind of the observer who rests upon it, and makes him less exact and trustworthy, while it may be full of danger to the patient.

These considerations alone would justify an attempt to render the prognosis of structural diseases of the heart more definite, but a study of these affections from the point of view of prognosis also leads to closer observation, and to the recognition of the importance of details which do not force themselves upon the attention so long as diagnosis in the ordinary and limited sense of the word only is the object.

A further justification of the prominence given to prognosis is that the grasp of all the facts of the disease and of the individual case, which is necessary to the formation of a just forecast of the result, is the best guide for treatment, whether this may demand chiefly patience and caution, or must be energetic and prompt. Prognosis is not merely a well-instructed conjecture as to the ultimate issue, it is a deliberate judgment as to the processes and tendencies of the disease, and as to the constitutional soundness and strength of the patient. The foresight relates to the dangers which attend the attack, to the course it will run, and to the influences and contingencies which make for or against the sufferer. To this there has only to be added a knowledge of the therapeutic measures by which the tendency to death or structural injury can be antagonized, and by which the patient can be guided and helped, together with skill, courage, and promptitude in applying it, and we have all the requisites for successful treatment.

Treatment, therefore, will form a natural sequel and corollary to prognosis.

HYPERTROPHY.

This condition of the heart will not detain us long. The prognosis of cardiac hypertrophy, like the symptoms, is that of its cause, and the character and degree of hypertrophy are important, not so much in themselves or on account of danger likely to arise out of them, but as indicating the existence of some condition which has given rise to the increase in thickness and contractile power of the muscular walls of the heart, and which is serious, possibly, in proportion to the hypertrophy which it has provoked.

Causes of Hypertrophy of the Left Ventricle.

The causes which give rise to hypertrophy must be enumerated. The left or right ventricle may be affected alone or predominantly, or both may have undergone change. The causes which bring about hypertrophy of the left ventricle are, in the first place, valvular diseases, stenosis, or insufficiency of the aortic valve, or mitral insufficiency. With these we are not now concerned. Next in frequency will be protracted high arterial tension. When high pressure in the systemic arteries is recognized, and especially when there are evidences of its having been habitual for some time, such as large, thickened, tortuous, radial and temporal arteries, or a dilated ascending aorta, the underlying condition to which it is due must be identified, and according as this is found to be renal disease more or less advanced, or gout, incipient or confirmed, or one or other of the affections, organic or functional, which are capable of inducing high arterial tension, will be both prognosis and treatment.

Left ventricle hypertrophy, again, may in rare instances have no other assignable origin than adherent pericardium, a condition which is difficult to diagnose, and with regard to which prognosis is extremely indefinite. Adherent pericardium may, however, be a factor in the causation of sudden death, and its recognition, therefore, is a matter of interest.

If hypertrophy is at any time produced by a mode of life entailing sustained muscular exertion, it is physiological and not pathological, and has no claims on our attention; unless, indeed, the initial change has been dilatation caused by violent and sustained effort, such as racing, the hypertrophy being secondary to this; but even here it is the dilatation and not the compensatory process which has to be considered.

Causes of Hypertrophy of the Right Ventricle.

Hypertrophy of the right ventricle, as of the left, has its most common cause in valvular disease—in this instance chiefly mitral stenosis or insufficiency;—and after valvular disease, in conditions which give rise to obstruction in the pulmonary circulation.

Emphysema and bronchitis are the lung affections which most frequently give rise to hypertrophy of the right ventricle, but the hypertrophy is rarely dissociated from dilatation, which is the primary effect of the obstruction in the pulmonary capillaries. Collapse of a portion of the lungs, contraction of a lung from pleural adhesions, fibroid phthisis, or any condition which throws a considerable area of lung surface out of gear, will in a certain degree give rise to overwork and consequent hypertrophy of the right ventricle.

Diagnosis of Hypertrophy of the Left Ventricle. Physical Signs.

The diagnosis of cardiac hypertrophy must be briefly described, if only for the purpose of pointing out the distinctions between it and dilatation. Taking, first, hypertrophy of the left ventricle: as regards the pulse, it is that of the condition which has given rise to the hypertrophy. When this has been arterio-capillary obstruction from renal disease, gout, lead poisoning, pregnancy, or other cause, the pulse will be that of high tension. The artery will be full between the beats, will not be easily flattened under the fingers, and can be followed some distance up the forearm like a cord. The pulse wave will not have a violent ictus, but will rise gradually and subside slowly, and as it makes little impression on the fingers it is sometimes described as weak, but when the attempt is made to arrest the wave very firm pressure is required. The description given applies strictly only when the artery is contracted and small, or of moderate size; when it is large, whether from physiological relaxation of its muscular coat or from distension by protracted high pressure within it, while the vessel can still be felt very distinctly between the beats, the pulsation is more conspicuous and abrupt.

The Heart.—On inspection, there may or may not be recognizable bulging over the cardiac area; the impulse is not extensive or violent in uncomplicated cases; the apex beat, if visible, is seen below the normal point, frequently in the sixth, sometimes even in the seventh, space, and is probably also displaced somewhat outwards; it is a circumscribed gentle heave. Occasionally one or more of the intercostal spaces above and to the inner side of the apex line may be seen to be retracted, but this is very rare, and is never so well marked in hypertrophy without valvular disease as it may be seen in aortic regurgitation.

By **palpation**, which is always an extremely important part of the physical examination of the heart, the apex beat is further defined, and is felt as a powerful but deliberate thrust in the space, sometimes distinctly lifting the adjacent ribs. The more the fingers are pressed into the space, the more distinctly is the thrust recognized.

When the flat of the hand is laid over the cardiac region, a general heaving impulse can usually be felt, but when the left ventricle is alone or mainly affected, it is not very conspicuous. As a rule, no impulse is felt to the right of the sternum except when the aorta is dilated, in which case pulsation may be felt when the fingers are thrust into the second and third spaces near the sternum. It must be added that sometimes neither apex beat nor impulse can be seen or felt when the hypertrophy is considerable. Mostly this is in deep-chested individuals with large, overlapping lungs, but occasionally hypertrophic enlargement takes a direction which carries the heart away from the chest wall.

Percussion maps out more or less accurately the enlargement of the heart downwards and to the left. This demarcation should be done with extreme care, but it must not be taken for granted that the outline drawn on the surface corresponds exactly with that of the organ, or gives a trustworthy idea of its size. To say nothing of the difficulty of defining deep dulness, the heart may enlarge backwards instead of laterally. The results of percussion must be correlated with all the other evidence as to the size of the heart.

Auscultation, besides teaching the character and intensity of the sounds, must be made to contribute to the information on this point by careful noting of the seat of maximum intensity of the sounds in the apex region.

The left ventricle first sound as heard at the apex is less distinct than in a normal state of the heart; either the mass of muscle enters into contraction less simultaneously, and the muscular tension being less sudden yields a duller sound, or the thickness of the walls masks the sound produced by the sudden tension of the valves and tendinous cords. The second sound, which at the apex is aortic only, is, on the contrary, usually louder than normal, and is often heard at or near the apex more distinctly even than in the right second space.

When the first sound is prolonged and muffled, and especially when it can be described as "impure" (a very objectionable term) or murmurish, careful examination will usually reveal that it is reduplicated, the two first sounds, produced by the right and left ventricles respectively, not coinciding. This indicates that the left ventricle is no longer quite equal to the extra work imposed upon it, and marks the supervention of a tendency to dilatation.

At the base of the heart the left ventricle first sound is still less distinct than at the apex, and is, indeed, frequently quite inaudible, while the accentuation of the second sound is rendered more evident by the absence of the first. As has already been said, however, the aortic second sound may be even more distinct at the apex than in the aortic area. When this sound is not only accentuated but low pitched and ringing, the root of the aorta is more or less dilated, and the sound will be heard for some distance to the right of the edge of the sternum, perhaps over a great part of the chest and along the spine.

The sounds of the right ventricle undergo no modification of sufficient importance to require notice.

Symptoms.

Various symptoms are described as resulting from hypertrophy of the heart—discomfort from the violence of the impulse, or actual pain in the region of the heart, tenderness on pressure in the neighbourhood of the apex, throbbing sensations in the head and neck, pulsatile noises in the ears, or audible pulsation in the carotid and other arteries. The action of the heart may be unduly frequent, or too easily excited, or abrupt and irritable, or irregular with falterings and bounds, which are very disturbing to the subject, and the heart may be prone to palpitation. There may be a sense of respiratory oppression, with sighing and desire to fill the chest with air.

Some of the symptoms are simply the result of the size of the heart and of the vehemence of its beat; others are due, not to the hypertrophy itself, but to its cause, whether in the valves or in the vessels, or to external influences which have given rise to it; others, again, are common to various affections of the heart, functional or organic, besides hypertrophy. They have no such direct or definite bearing on prognosis as would warrant a discussion of their significance, though a sustained frequency of the pulse is an unfavourable sign.

PROGNOSIS.

The question of prognosis in relation to hypertrophy mostly resolves itself into this: whether the compensation which it establishes is adequate and efficient, and how far it promises to be durable. The danger that the hypertrophy may go beyond the requirements of the occasion which has called for it does not, in my opinion, need consideration.

Compensation is efficient when there are no symptoms of embarrassment of the circulation, and when the heart responds to all ordinary calls upon it without undue shortness of breath or respiratory distress. The effects of exertion are an important criterion, due allowance being made for the greater liability to breathlessness which is natural to some individuals, or is produced by bodily conformation or which results from a sedentary mode of life.

But the sounds of the heart usually give notice when it is overtaxed by the resistance to the onward movement of the blood. The interval between the first and second sound may be prolonged, the systole requiring more time than under ordinary circumstances to complete itself. So long as the normal proportion between the systolic and diastolic pauses is not disturbed, there is no indication that the heart is unequal to its work or is suffering from the stress put upon it: but when the systolic interval is lengthened at the expense of the diastolic, so that the sounds are equidistant, the period of repose and reconstitution of the muscular fibres of the ventricle is shortened, and their nutrition must in time suffer. During systole the blood is squeezed out of the walls of the heart, and it is during the diastolic relaxation that it obtains free access to the cardiac fibres.

Another evidence that the heart is yielding to the strain of overwork is reduplication of the first sound. That is, as has been already said, due to want of synchronism between the two ventricles in the act of contraction, or rather in arriving at that point in their contraction when their valves and tendinous cords and muscular walls are all made tense. This reduplication is at first very slight, the two sounds are separated by the briefest possible interval, and are distinct one from the other only at one spot, just to the inner side of the apex; elsewhere there is only a slurring or prolongation of the first sound. Later the sounds of the right and left ventricles are quite distinct, and the duplex first sound is recognized over a considerable area, usually in the direction of the ensiform cartilage, but sometimes upwards. One variety of the cantering rhythm or *bruit de galop* is produced by this doubling of the first sound, where the successive sounds are one—one, two. Sometimes the dislocation of the ventricular first sound is so considerable that it is a task of extreme difficulty to identify the sounds of the heart at all, and associate them with the systole or diastole respectively so as to say which is first and which second. The most striking and confusing examples are met with in aortic stenosis and pericarditis, but they occur also in high arterial tension. The prognosis becomes serious when the first sound is broken up in any very considerable degree.

But when the cardiac hypertrophy has been brought about by high arterial tension, whether associated with renal disease, or gout, or other cause, there must be taken into consideration the possibility that the powerful heart may rupture diseased vessels in the brain, and if the arteries are conspicuously degenerated it is better for the patient that the heart also should undergo some degree of degeneration, and fortunately this usually takes place.

Treatment.

Treatment for hypertrophy, as such, has always appeared to me to be out of place. The functional vigour and energy of the overgrown and overstrong heart could no doubt be reduced by various means—low diet, enforced rest, and such drugs as aconite—but unless it is clear that the hypertrophy has gone beyond the requirements of the condition which has given rise to it, the advantage of this procedure would be more than doubtful. Even in the attempt to relieve the incidents of hypertrophy, palpitation, throbbing sensations in the chest, præcordial oppression, the employment of direct cardiac depressants is rarely of service, and is at times attended with danger. In aortic stenosis, for example, I have known aconite, given with a view of quieting tumultuous action of the heart, so far reduce the contractile energy of the left ventricle that it was no longer able to cope with the obstruction, and death quickly followed from cardiac asthenia, the pulse becoming imperceptible, the extremities livid, and the surface of the body cold and damp. This would be less likely to occur where the cause of the hypertrophy was obstruction in the peripheral circulation, as the arteries and capillaries are relaxed by such agents as depress the action of the heart.

The treatment of the chief causes of hypertrophy, aortic disease, and high arterial tension have already been discussed elsewhere, so that what we have to consider here is not the treatment of hypertrophy or its causes, but treatment suggested by the hypertrophy. We shall recognize, for example, the necessity of diminishing the volume and improving the quality of the blood by appropriate diet and hygiene, and, if necessary, by tonics. We shall recognize, also, the desirability of diminishing the resistance to the onward movement of the blood in the arterio-capillary network by care in diet again, by aperients, and by eliminants of various kinds. In some cases, the resistance in the peripheral circulation may be further lessened with advantage by the physiological relaxants of the arterioles and capillaries, such as nitroglycerine and the nitrites. By these means the work thrown upon the heart is reduced, and, if necessary, the heart may be strengthened by such remedies as strychnine and digitalis.

All these measures are specially required when the hypertrophy is no longer quite equal to the task which it had originally been developed to perform, and reduplication and other modifications of the sounds are present. It is now that the incidents of hypertrophy, palpitation and the like, are most commonly complained of, and they will be best alleviated or removed by the measures just sketched, by relieving the heart of work on the one hand, and by helping it on the other.

DILATATION.

The word "dilatation" requires no explanation as applied to the cavities of the heart. Individual cavities may be dilated, as, for example, the left auricle in mitral stenosis or regurgitation, the right ventricle in pulmonary emphysema or mitral disease; but when spoken of as a form of heart disease, dilatation usually means dilatation of the left ventricle, mostly with, but sometimes without, dilatation of the right ventricle. Together with the expansion of the cavity of the ventricle there may be more or less thickening of its walls, representing an attempt at hypertrophy, compensatory of the dilatation itself, or of the difficulty in the circulation which has led to it. Not unfrequently the hypertrophy has preceded the dilatation. The walls of the heart may have an approximately normal thickness, which will really imply a certain degree of hypertrophy, or they may be distinctly thinned. The parietes may have a normal colour, consistence, and structure, or they may be pale, flabby, and degenerated, or dense and tough from fibroid substitution. But even more important than the anatomical condition is the physiological or functional condition, the special characteristic of which is that the ventricle does not complete its systole, but only expels a portion of its contents.

Dilatation of the heart may be contrasted with hypertrophy rather than compared with it. Hypertrophy is compensatory and an evidence of vigour; dilatation is, for the most part, a confession of failure on the part of the heart muscle and an aggravation of other causes of interference with the circulation which may be in operation. In hypertrophy the augmented mass and increased strength of the muscular walls enable the ventricles to complete their contraction in the face of difficulty. The result of dilatation is that the ventricles habitually fail to expel the whole of their contents. Very frequently it is only a very small proportion of the blood which is projected into the great arteries. In well-marked cases the chambers of the heart are always full, and little blood being received and expelled, there is a stagnation in the auricles and ventricles which may allow of the deposition of fibrin among the fleshy columns and pectinate muscles. It has seemed to me that the imperfect emptying of the ventricles has not always been fully realized as the special feature of dilatation, but it will be seen that if a dilated ventricle launched the whole of its contents into the arterial circulation, the amount being much larger, the rate of movement of blood would be greatly accelerated, whereas the contrary is the case.

A distinction must be drawn between the dilatation due to inherent weakness of the muscular walls of the heart and that attending valvular disease.

The dilatation attending aortic regurgitation belongs to a different category from primary dilatation, as has already been shown, and, instead of aggravating the difficulties of the circulation, it is a part of the compensatory arrangement. It is clear that if a certain proportion of the blood projected into the aorta at each systole returns to the ventricle during diastole, it is an advantage, and indeed a necessity, that the capacity of the ventricle should be increased, so that, in spite of the reflux, a normal amount of blood may remain in the arteries and be distributed to the tissues. The dilatation, however, in aortic regurgitation is accompanied by hypertrophy which enables the ventricle to contract perfectly, so that the characteristic of dilatation—the partial and imperfect emptying of the ventricle—is not present.

It is probable, again, that the dilatation of the left ventricle met with in mitral regurgitation may have a similar compensatory effect. It would seem to be an advantage, since some of the blood regurgitates into the auricle, that the ventricle should contain sufficient to allow for this, and yet discharge a due amount into the aorta. In dilatation consecutive to mitral regurgitation, moreover, there is a certain degree of hypertrophy, and the systole is carried through.

The dilatation of mitral regurgitation stands, then, on a totally different footing from primary dilatation.

The mode of production of dilatation of the heart is highly complex; it is usually understood to be the result of a gradual yielding of the walls of the ventricles, either from their own inherent weakness or from undue resistance to the onward course of the blood. It is often supposed also, when dilatation exists, that it is an established and more or less unvarying or progressive condition. Both these ideas require modification.

All violent or protracted exertion is attended with temporary dilatation of the heart, which may go so far, even in strong and healthy persons, as to give rise to temporary murmurs. Loud murmurs so produced may be heard at the apex, over the tricuspid area and over the pulmonic area, in individuals who have no evidence of cardiac weakness at the time, and who develop no valvular or other cardiac disease for many years afterwards. A personal friend always had a loud, systolic, pulmonic murmur after hunting, which sometimes, when there had been a severe run, lasted two or three days. It is, again, not very uncommon for young and strong men to return from climbing in Switzerland with more or less dilatation of the heart, which may persist for weeks. This is usually when little exercise has been taken during the year, and considerable ascents or very long walking excursions have been made without sufficient preliminary training. Boat-races no doubt give rise to temporary dilatation, and it may be met with as a result of training for races. It might, perhaps, be better to speak of distension of the cavities of the heart in these instances, rather than of dilatation.

But the circumstances which are capable of producing a temporary distension of the ventricles in a sound and vigorous state of the organ will be competent to give rise to dilatation when it is weak and flabby, and when other conditions are present which tend to dilatation, and it is more likely that the weakly ventricles give way from time to time under stress and fail to recover perfectly, than that the yielding is gradual and continuous. In contracted granular kidney with persistent and often extreme arterial tension, dilatation at an early period is comparatively rare, which would scarcely be the case were continuous resistance to the emptying of the ventricle the most efficient cause.

In violent exercise the pulse in becoming extremely frequent also becomes extremely soft and short; the arterioles and capillaries are relaxed in order to facilitate the rapid movement of the blood which is necessary to supply fuel and oxygen to the muscles during exertion, and the arterial tension is low. The resistance to be overcome by the ventricles is thus reduced to a minimum, which diminishes the liability to over-distension.

In the production of dilatation a common and important, if not a constant, factor is habitual high arterial tension. A second factor, less constant perhaps but more important, is inherent weakness in the ventricular walls. Sometimes one will predominate, sometimes the other. With these proximate causes, to use an old-fashioned term, will co-operate very varied influences—exertion, excitement, chills, imprudence in eating and drinking, constipation, by augmenting the stress on the heart, a sedentary mode of life, anæmia, anxiety and depressing emotions, by diminishing its power of resistance.

Acute dilatation of the heart is more common than is generally supposed.

In a very large proportion of the cases admitted into hospital suffering from symptoms due to cardiac dilatation, there has been an acute aggravation of the affection from work or exposure, and under treatment a considerable diminution in the size and capacity of the heart is commonly observed. But cases occur in which there is ground for supposing that dilatation has been induced at once in a heart not previously affected. The following are examples selected as illustrations.

A gentleman at the age of 70, in vigorous health and capable of any ordinary amount of exercise, overtook a labourer pushing a heavily laden wheelbarrow uphill, who had to stop and rest every few yards. Proud of his strength, he told the man to stand aside, and himself wheeled the barrow for some distance at a good pace. He lost his breath and found that he did not recover it as he expected, but that he continued to pant and to be conscious of violent action of the heart, accompanied by a sense of oppression in the chest. He got home with difficulty, the least exertion was attended with shortness of breath, and he could not rest at night. After a few days he sent for his medical man, when the physical signs of dilatation of the heart were found to be present. Mitral insufficiency was quickly established, and when I saw him considerable dropsy existed. He died shortly afterwards.

A gentleman, aged about 55, remarkably strong and active, who was said to have had a slight attack of pleurisy shortly before, ran quickly up a long flight of stairs in the City. On arriving at the top he was found gasping for breath, unable to speak and scarcely able to stand. He was soon sufficiently recovered to be sent home, and a few days later he was brought by his medical attendant to my consulting-room, chiefly in order that I might aid in enforcing the rest and care which were considered necessary. The patient admitted that he was weak and soon out of breath, but declared that he was quite equal to business. The pulse was irregular in force and frequency, the apex beat of the heart diffuse, devoid of the impulse or thrust, and displaced downwards to the sixth space and outwards beyond the nipple line, and the other physical signs of a considerable degree of dilatation were present. With great reluctance a certain amount of rest and treatment was submitted to, and the heart and pulse became stronger and steadier, and the apex beat came in towards its normal situation about an inch.

Two months later the patient ran down from a committee-room in the House of Commons for a bag of papers which he had forgotten, and back again. An attack of the same kind as that just described came on, and this time lasted longer. The evidences of dilatation of the left ventricle were more marked, and œdema of the ankles soon came on. The patient was kept in his room, and as far as practicable at rest in bed or on a couch. Improvement was again obtained, but much more slowly than before, and as my visits were the chief restraint upon him, after a brief attendance at short intervals, further consultations were put off indefinitely.

Within a few days the exertion of going downstairs and a serious imprudence in diet brought on a return of symptoms, and the œdema of the lower extremities was shortly complicated by thrombosis of the deep femoral vein of the left side, which extended along the iliac vein to the vena cava and down the right iliac and femoral veins. A tedious illness followed, which finally proved fatal by extension of the thrombosis to the renal veins, the heart itself appearing to improve somewhat.

Another case was that of a boy who was admitted into St. Mary's Hospital with extreme dilatation of both ventricles. He had had no previous illness, but had been underfed. Most Londoners will be familiar with the appearance of lads in the scanty attire of the cinder-path, who are making use of the streets as a training-ground for races, usually in the dusk of the evening, after working hours; perhaps, also, this time is chosen to avoid police interference. I have myself often been astonished at the pace and endurance of these athletes-under-difficulties. The patient had been training in this way, and had persevered in spite of shortness of breath, till severe symptoms set in. He died soon after admission, and the ventricles were found to be enormously dilated.

A certain degree of dilatation of the left ventricle usually occurs at the onset of acute renal disease, under the combined influence of the resistance in the peripheral circulation and of the enfeeblement of the heart. In other acute diseases this may happen. I have twice seen considerable dilatation of the heart in mumps attended with collapse and cerebral disturbance; in one case, it was fugitive; in the other, it lasted some time and recurred afterwards. In both, the arteries were tightened up, giving rise to high tension and great resistance to the onflow of the blood. Whether the severe collapse and delirium which sometimes supervene as the acute symptoms are subsiding in mumps are always associated with dilatation of the heart, it is not in my power to say.

A common cause of dilatation of the heart is anxiety. Nothing is more certain than the influence of prolonged mental depression upon the heart, and the sensation of aching, oppression, and weight which attends grief and anxiety, and which was considered to point to the heart as the seat of emotion generally, is indicative of an injurious effect upon this organ. It is almost literally true that people die of a broken heart. The combination of overwork, excitement, worry, and trouble, often met with in City life, especially on the Stock Exchange or in mercantile or financial circles during a commercial or financial crisis, brings us many cases of cardiac dilatation among men, and it is needless to say that domestic anxieties—grief on account of children who have died or given trouble—have the same effect among women.

Among the special causes of dilatation of the heart —acting, no doubt, on pre-existing tendencies—which have come under my notice, are injudicious hydropathic treatment, the so-called Banting method of reducing obesity, and inhalation of the fumes of Himrod's powder for the relief of asthma.

In one case a gentleman suffering from dyspeptic symptoms, probably due to cardiac weakness and consequent sluggish circulation in the abdominal viscera, underwent a routine treatment by baths, wet cold packs, and compresses, under which attacks of dyspnœa came on and œdema set in. He had previously been under the care of competent observers, who had not found any serious degree of dilatation. The dropsy advanced in spite of treatment, and when I saw the patient it was considerable, and the physical signs gave evidence of extreme dilatation and thinning of the left ventricle. Death occurred suddenly during a paroxysm of dyspnœa.

Another patient came straight to my consulting-room from a hydropathic institution, where he had undergone vigorous treatment for digestive and liver derangements, attributed to long residence in India. He was very anæmic, and breathless; the heart was greatly dilated and its action irregular, and there was incipient œdema. Fortunately he recovered.

One of the most extreme cases of dilatation I have ever met with was in the case of a lady who had undergone an amateur course of Banting treatment for obesity. She had lost some of her fat, but had become extremely breathless, so that to walk across a room or put on an article of clothing with the assistance of her maid caused her to pant and gasp for breath in a very painful way. The pulse was weak, small, and very irregular; no impulse or apex beat could be felt, and the size of the heart, as mapped out according to the deep dulness, was incredibly large, had not the results of percussion been confirmed by the sounds being audible over the entire area of dulness, and, later, by a feeble apex beat in the seventh space near the mid-axillary line. Apparently habitual high arterial tension had been exaggerated by an exclusively nitrogenous diet, and, under it, the ventricle had given way.

It may, perhaps, be well to add that in the treatment of obesity by beefsteak and copious draughts of hot water there has not been, according to my experience, increase of arterial tension, but the reverse.

The following case is the most serious of several attributable to Himrod's powder:—

In December, 1883, I saw, with Dr. Andrews, of Hampstead, a gentleman, aged about thirty-six, apparently of sturdy constitution, who, after an attack of bronchitis, suffered from several nocturnal attacks of asthmatic dyspnœa. A month at Hastings had apparently restored his health; but, after a fortnight of work in London, his nocturnal asthma and shortness of breath were as bad as ever. For the asthmatic attacks he had inhaled immoderately the fumes of Himrod's powder, obtaining relief for the time, but with disastrous after-effects. The whole house was reeking with the odour of the fumes at the time of our consultation. Dr. Andrews had witnessed the rapid development of the condition of the heart which existed. The area of cardiac dulness was greatly increased; no apex beat was recognizable; there was a doubtful systolic murmur (mitral) in the region of the apex, a loud systolic murmur (tricuspid) near the ensiform cartilage, and a faint systolic aortic murmur. Tricuspid incompetence was further distinctly manifested by great enlargement of the liver, and by marked jugular pulsation. There was nothing in the patient's habits, or mode of life, or previous history which was at all calculated to give rise to dilatation of the heart, and I had the less hesitation in attributing it to the solanaceous fumes, from having seen similar effects in other cases. In about a fortnight, after free purgation by calomel and the administration of digitalis, the patient was much better, free from asthma, and able to walk upstairs, while the liver had gone down to its usual size. The heart remained very large, gave no defined apex beat and only a diffuse general impulse, while a high-pitched mitral murmur, a louder tricuspid murmur of low pitch, and a faint aortic murmur, all systolic, were audible. The action of the heart was curious; now and then there was a sudden bump against the palm of the hand placed over the right ventricle, and it was found that the beats of the heart were in pairs, only the first of which was accompanied by a systolic apex murmur, the second having a loud first and second sound, but scarcely reaching the pulse. With both beats the loud tricuspid murmur was present. The improvement continued, so that the patient resumed his duties before the end of January, and attended regularly to business through the month of February. In March came another relapse which led to dropsy, and the patient died early in April. Fortunately an examination of the body was permitted, notes of which are as follows:—

There was much œdema of the legs and right arm, but no fluid in the abdomen. The heart was enormous, measuring six inches and a half from base to apex, and sixteen and a half in circumference. It was flabby, lying flat, pale, fatty looking, not lacerable. The right auricle was greatly dilated and very thin, the wall being translucent at one part; the appendix was filled up by solid clot and stained black by blood. The tricuspid orifice took four fingers easily; there was no roughness. The right ventricle was enormously dilated, and would almost have held a duck's egg; the walls were thin, soft, and flabby, the valve stained black; the flaps were thin and the cords delicate. The pulmonary artery and valves were normal. The left auricle was not at all dilated, the appendix contrasting with the right, and compared with the other cavities curiously small. The mitral orifice would admit two fingers. The left ventricle, like the right, was enormously dilated, and looked as if it would hold a fist: the walls were thin, mottled, and flabby, but not lacerable; the papillary muscles were small, the mitral valve and cords quite normal, and remarkably free from thickening. The aorta was very small and the valves normal.

Diagnosis.

The diagnosis of dilatation of the heart is comparatively easy to any one who has had a fair amount of experience, and is prepared to exercise sufficient care in making the requisite physical examination. Certain precautions are necessary which will be pointed out later. When, however, the dimensions of the heart have been made out, and an estimate has been made of the relative degree of dilatation and hypertrophy, and of the comparative condition of the two ventricles, a small step only has been taken towards the attainment of the knowledge which is necessary in order to forecast the course and issue of the case, and to direct the treatment. This may be illustrated by a reference to experience which must have been met with by most physicians. A patient is seen on account of dropsy and other severe symptoms due to dilatation of the heart. Under treatment he gradually recovers, and is apparently much in the same state of health as before. After a shorter or longer time—months in some cases, years in others—a relapse of the symptoms indicative of cardiac failure takes place. The general conditions may not apparently be worse, there may be no marked difference in the physical signs, or in the state of the pulse; but the patient's chances of recovery are not the same. Besides the fact that he is older by so much, a change has taken place in the muscular structure of the heart which is not appreciable by means of physical signs. There is not the same reserve of force or nutritional vigour, and the response to treatment is not the same. If a second attack does not prove fatal, a third or fourth will, and the briefer the interval, the more serious is the significance of the relapse, unless it is accounted for by some sufficient cause. But the defective vital energy, which in one case becomes manifest only after one or more serious derangements of the balance of the circulation with dropsy and other symptoms, may in another be present from the first. There may be hereditary tendency to weakness of the heart, or the heart may be worn out by an unhealthy mode of life, or by protracted emotional strain; or the patient's tissues generally may lose their nutritional vigour early. Facts of this class are not revealed by physical signs; some of them may be known to the family medical man, but otherwise they are ascertained only by persevering and careful inquiry. It is obvious, then, that many other considerations enter into a true and thorough diagnosis of dilatation of the heart besides its anatomical condition.

With this introduction, the physical signs which reveal dilatation may be described together with the pulse.

Pulse and Physical Signs.

The **pulse** in advanced dilatation of the heart is usually irregular in rhythm and unequal in force of beat, and is sudden, short, unsustained, and usually easily compressed. The artery at the wrist may be large or small; it will be specially large when there has been antecedent high tension which has dilated the arteries. The significance of a pulse of this character is not absolute, as all these characters may be present when there is no recognizable change in the walls, cavities, or valves of the heart, apparently from disordered nervous influence only. The pulse, again, need not be irregular in advanced dilatation of the heart so long as the patient is in repose and the breathing is tranquil and easy. The regularity, however, is easily disturbed by exertion or effort, or by bronchitis, or merely by deep breathing. In moderate and slight dilatation the pulse may be regular, but, while irregularity is the rule, there is no such constant relation between the degree of regularity or irregularity of the pulse and the amount of dilatation of the heart as to make one diagnostic of the other.

The Heart.—In the examination of the heart, inspection and particularly palpation will be of the greatest importance. The visual examination will be directed to the situation and character of the apex beat and to the impulse of the right ventricle. Retraction or bulging of the intercostal spaces, elsewhere than at the visible apex beat, will be noted.

The apex will be displaced outwards and downwards, and it may be visible over a large area. Very commonly it cannot be seen at all, and the point of maximum apparent impulse does not necessarily belong to the true apex. The right ventricle impulse will be diffuse, if recognizable at all, and as a rule it is inconspicuous.

Palpation in most cases of dilatation furnishes information which contributes more to precision of idea as to the actual state of the heart than any other branch of physical exploration. The right hand should be applied closely over the entire cardiac region, the palm over the right ventricle, the fingers, spread out and close together alternately, over the apex region. Distinct impulse over the right ventricle, while it indicates more or less obstruction in the pulmonary circulation, indicates also some degree of vigour in this ventricle available for compensatory work. A mere vibration has a converse significance. The first object of attention, however, will be the identification of the point of maximum impulse in the apex region, and a careful estimation of the area over which the apex beat extends, and of its force and character; whether, for example, it is a mere concussion of the chest wall or a more or less distinct thrust at any point. Sometimes the word "slapping" is employed to describe the impulse or apex beat characteristic of dilatation. Further exploration will be made by pressing the fingers into the intercostal spaces around and beyond the point of maximum impulse, and it must be borne in mind that this impulse may not be the real apex beat, but the impulse of the right ventricle. Sometimes impulse is detected much above the normal situation, in the fourth space perhaps as far outwards as the anterior axillary fold or behind it, or it may be concealed by the female mamma. It may be the left edge of the right ventricle resting on the interventricular septum which is here felt, or a part of the rounded apex of the left ventricle. According as the apex is capable of giving a distinct thrust or communicates only a diffuse shock, and, according as the beat is well-defined and steady, or vaguely felt over a considerable area, will be the estimate of the degree of dilatation and of the thickness or thinness of the heart wall. Not unfrequently neither impulse nor apex beat can be detected, or the impulse is so vague that it cannot be localized; unless this is due to overlapping lung, it indicates great weakness of the muscular walls of the heart.

By **deep percussion** the outline of the heart can be more or less accurately mapped out. It is more rounded in the apex region than normal, and the area of dulness is greatly extended to the left. When no impulse of any kind can be felt we may have to depend entirely on percussion for information as to the size of the heart. I have found continuous deep dulness outwards as far as the mid-axillary line and downwards to the seventh space, shown to be cardiac by the intensity of the sounds at the extreme limits, and, when the heart had gained strength, by an apex beat recognizable by palpation at the farthest point.

Auscultation.—The characteristic modification of the

sounds of the heart produced by dilatation is that the left ventricle first sound becomes short; usually it is also louder than normal. Probably from this change in the character of the left ventricle first sound it is almost always audible in the aortic area, contrasting in this respect with hypertrophy. It is audible also to the left of the apex beat. When no impulse or apex beat can be felt, the sounds, and especially the first, must be made use of to ascertain how far to the left, the left border and apex of the heart have been carried by the dilatation, and in what degree the enlargement is due to dilatation or hypertrophy. Percussion, of course, maps out the deep dulness, and shows approximately the limit of the heart and its extension to the left, but the point of maximum intensity of the sounds and the area over which they are audible will corroborate or correct the idea formed as to the size of the heart from percussion, and percussion yields no information whatever as to the kind of enlargement, but leaves it to be supplied by the character of the sounds on auscultation.

Not uncommonly dilatation of the left ventricle is accompanied by and gives rise to a systolic apex murmur, obviously due to mitral incompetence. This is induced by imperfect accomplishment of the constriction of the orifice, which is part of the normal contraction of the ventricle, and which co-operates with the curtains of the valves in preventing regurgitation into the auricle. Now mitral valvular disease is attended with precisely the same combination of conditions—incompetence of the valve and dilatation of the ventricle—and of physical signs—a systolic murmur and displacement of the apex beat with increased cardiac dulness. The prognosis is very different in the two cases, and it is therefore extremely important to distinguish between them. This cannot be done with any degree of confidence by means of physical signs alone, but it may be said that the presence of the left ventricle first sound in primary dilatation and its absence in primary valve lesion will sometimes be of great service as a factor. The age of the patient and the history of the case will be of great help, and the mode of onset of the symptoms and their causation should be carefully inquired into. If the patient is middle-aged and the symptoms date from some imprudent over-exertion, the presumption will be in favour of primary dilatation. If there is a history of acute rheumatism, there will be little doubt as to the diagnosis, especially if the patient be a child or young adolescent.

There is nothing in dilatation of the cavities of the heart to affect specially and directly the second sound, but, in proportion as this condition of the ventricles impairs the propulsive energy of the systole the second sound will be enfeebled, and the aortic second sound is usually weak as compared with the left ventricle first sound. The relative loudness of the sounds therefore enters into the considerations from which may be calculated the efficiency of the ventricle.

Very important information is often supplied by observation of the intervals between the first and second and the second and first sounds, the short and long pauses respectively. When, with dilatation of the left ventricle there is resistance in the arterioles and capillaries, which is very commonly the case, the interval between the first and second sound may be prolonged, and as this marks the duration of the systole it shows that the ventricle is endeavouring to cope with the difficulty and to complete its contraction. The short or systolic pause may be so prolonged as to equal the diastolic or long pause, and the sounds thus become equidistant, the first also having become short and sharp; the only difference between the sounds is one of emphasis or of pitch, and it is often difficult to say which is which. The sounds may be compared, when the heart is acting slowly, to the ticking of a clock; when rapidly, to the puffing of a distant locomotive.

On the other hand, the first and second sounds may be approximated, and as we cannot suppose that a dilated and enfeebled ventricle completes its systole in a shorter time than normal, the only possible explanation is, that it is quickly brought up short by the resistance in the arterial system and expels but a small proportion of its contents. This abbreviation of the systolic pause is therefore a serious indication of failure of the ventricles, and when carried to an extreme, so that the second sound follows the first immediately and almost seems to overtake it, is significant of immediate danger.

In two successive attacks of symptoms due to dilatation of the heart, when the degree of dilatation appears to be exactly the same, a difference in the length of the systolic interval is sometimes the chief, if not the only, point which makes a difference in the prognosis. The first and second sounds are spaced in the first attack, which is survived, and approximated in the second, which proves fatal.

Symptoms.

The symptoms which attend the earlier stages of dilatation are extremely varied and very vague. An imperfect and fluctuating supply of blood to the brain will give rise to impairment of bodily and mental energy, and to irresolution and vacillation of purpose; the memory is liable to fail, especially in regard to recent events, and the power of sustained attention and consecutive thought is diminished. The frame of mind will often be despondent, and the temper may be irritable. There may be attacks of giddiness or faintness; in one case, the patient, a vigorous old gentleman, aged 78, fell down from time to time as if he had been shot, with momentary loss of consciousness, getting up again at once, apparently none the worse. Sleep is usually disturbed, and sometimes almost absent, and, whether the nights are good or bad, there may be slumber at any period of the day, the patient dropping off to sleep even over his morning newspaper. For the same reason, viz. the irregular and imperfect blood supply and the back pressure in the veins, digestion and the action of the liver will be deranged. The appetite is bad, and food is followed by discomfort and by flatulent distension, which latter again reacts on the heart, and gives rise to oppression, or palpitation, or irregular action. The bowels are usually torpid. The urine is deficient in amount and high coloured, and there is often an habitual deposit of urates. Turbidity of the urine day after day, whatever the food and drink, or weather, or mode of life, should direct attention to the state of the circulation.

There is no one, and scarcely any combination, of the symptoms enumerated, however, which may not occur independently of weakness and dilatation of the heart, especially in states of system attended with high tension; and it would be waste of time to attempt to disentangle those directly due to the state of the heart from those which are merely accidentally associated with it.

The special symptom which calls attention to the heart as the probable starting-point of a number of ailments is breathlessness on slight exertion, but even this may be produced by other causes—by anæmia, for example; after middle age, pernicious anæmia may often give rise to extreme breathlessness, which may excite a suspicion of heart disease. High arterial tension alone, which has not yet given rise to dilatation, simple debility, very sedentary habits, may also cause great shortness of breath. These facts are mentioned in order to warn against a too ready inference that heart disease exists simply because the breath is short. An interesting fact in connection with breathlessness due to dilatation of the heart is that it is often relieved by exercise of the voice. I have met with several instances in which a clergyman has climbed into the pulpit with the utmost difficulty, and has not only preached a sermon comfortably, but has been all the better for it. A sense of breathlessness coming on during repose, and inciting the patient to make frequent deep inspirations, is usually a symptom of nervous depression, and has no necessary relation to heart disease. For the most part, as the disease advances, symptoms arise which are indicative of back pressure in the systemic veins, a gradually advancing œdema, and the like.

In an extreme case the patient will be dropsical, œdema invading the thighs, loins, and abdominal parietes, as well as the legs, and there may be fluid in the abdomen, and perhaps in one or both pleural cavities. The feet and legs will be cold and pale, or purple and livid, especially if hanging down; the hands also will be cold, and are often crimson or purplish, and the nails of a deep or dusky tint, instead of a bright pink. A white patch on the hand, produced by pressure, is very slowly invaded by returning colour. The sufferer is probably unable to lie down in bed, and is propped up by pillows, or he must have his legs down, and therefore spends day and night seated in his chair. Remarkable exceptions, however, are met with in this respect, some sufferers, while extremely ill, being able to lie down without distress. The face is pale, and perhaps puffy, especially under the eyes, with injected capillaries over the cheeks, and wears an expression of distress, and the eyes are watery. The lips are pale or bluish; the breathing is more or less laboured, even in repose, and the sufferer constantly supplements his reflex respiration by voluntary deep breaths. When he speaks it is in fragmentary sentences, and with evident effort and aggravation of the respiratory distress. The least movement brings on shortness of breath, which is often painful, even to witness. The pulse is frequent, irregular, and probably greatly deficient in tension, but not always. The urine will be scanty, of a deep colour, and probably high specific gravity; it most commonly throws down a copious deposit of brickdust lithates, and it may contain albumen, the amount of which may vary from day to day. The appetite will be very bad, and there may be nausea; the tongue probably furred, the bowels constipated or irregular. One of the most distressing symptoms is sleeplessness, and when, after hours of weary shifting of position, the sufferer is overcome by fatigue and drops off, he has painful dreams, and wakes suddenly in affright and suffocation for lack of voluntary help to his respiration. The longest and best sleep will be obtained while sitting in a chair, and sometimes by day rather than in the night.

Pursuing the examination, the jugulars will be found full, but not, as a rule, greatly distended or pulsating forcibly. The liver will be enlarged, coming down sometimes as low as the umbilicus, and extending across the epigastrium into the left hypochondrium. It will often be jogged by the right ventricle, but does not usually exhibit true expansile pulsation. The jugular distension and pulsation, as was pointed out to me by Dr. Mackenzie of Burnley, may be greatly exaggerated by pressure on the enlarged liver. There may be fluid in the peritoneal or pleural cavities, more commonly not, and there will, in most cases, be physical signs of œdema and congestion of the lungs; occasionally the percussion note may be good and the entry of air free and unattended with adventitious sounds of any kind to the very base of both lungs.

Prognosis when the Symptoms are Severe.

We may consider first the most serious case where the symptoms present are such as have just been described. We are called upon to answer the question, Has the patient a chance of recovering from the condition described, or will he die?

The first element in the judgment to be formed will be the urgency of the symptoms, and special importance will attach to two of the series—the nausea and loss of appetite and the sleeplessness—from the effects which they have on the patient's strength. Frequent vomiting of food is of very grave import, not only because the patient does not get the benefit of the nourishment, but because it shows that the stasis in the abdominal circulation has reached a point which interferes seriously with the digestive secretions. Attacks of faintness and of extreme exhaustion, or of severe dyspnœa, are also very serious.

It is not always when the dropsy is excessive that the condition of the patient is worst. From the late appearance or entire absence of dropsy in fatty degeneration of the heart, in aortic disease, and mitral stenosis, it would appear that a certain degree of pressure in the arterial system is required to co-operate with the back pressure in the venous system for the full development of dropsical effusion; and when the œdema remains moderate in amount, while other symptoms—such as breathlessness, faintness, and muscular weakness, the latter especially—indicate great cardiac inefficiency, it may be because the left ventricle propels the blood with very little force. Degeneration may be a factor in the case. A frequent, short, unsustained pulse and heart-beat, without much œdema, may thus indicate a more grave condition than extreme dropsy. Long-continued excessive frequency of the pulse is an unfavourable sign.

The urgency of the symptoms being about the same, a larger degree of dilatation (as indicated by a more extended area of dulness, displacement, and especially by greater diffuseness and weakness of the apex beat and impulse and greater shortness of the first and weakness of the second sound) will add gravity to the prognosis, and it will be borne in mind that approximation of the first and second sounds is always a very serious indication. Favourable points in the physical examination will be a good right ventricle impulse, and it has appeared to me that there is a greater margin for treatment when the liver is considerably enlarged.

When the symptoms and physical signs have been well weighed, the first question to be asked in view of prognosis is whether there has been any adequate exciting cause of the symptoms. If there has been recognizable over-exertion or excitement, or grave anxiety—any mental or bodily strain—or if there has been a chill, giving rise to bronchitis or other affection of the lungs, or deranging seriously the liver and digestive organs, it may be hoped by means of rest and treatment to undo the ill-effects and restore the balance. If, on the other hand, the symptoms have crept on gradually without traceable cause of the kind mentioned, and especially when due care has been exercised, and there has been no habitual error of regimen or neglect of bowels, the probabilities are that the symptoms are the outcome of conditions which cannot be reversed—of radical inherent weakness of the heart.

The previous mode of life, active or sedentary, careful or imprudent, and especially the habits with regard to alcoholic stimulants, will have a very important bearing on the probable issue of the attack, as will also the general soundness and the absence or existence of disease of any other important organs, especially of the kidneys and liver. The patient's cheerfulness, hopefulness, and courage under his sufferings, or his despondency, will make powerfully for or against him, not only as direct influences, but because the state of mind is often an index of the state of the system.

Another inquiry of great prognostic weight will be as

to the patient's family history. It is through this that we obtain an idea of his vital tenacity and of the trustworthiness of his tissues, and sometimes of the special liabilities of his heart. There are few tendencies which run more strongly in families than those which are manifested in the heart and vascular system, whether to high arterial tension with the effects on the vessels and heart, which follow from this, or to dilatation and weakness, or to degeneration, and a prognosis, otherwise not unfavourable, might have to be instantly revised on learning that the father and an uncle or two, or a brother, had died at about the patient's age from heart disease. Finally, the response to treatment will speedily afford an indication of the utmost value.

Treatment.

In the treatment of advanced symptoms due to dilatation of the heart, we have to deal at the same time with defective propulsive power on the part of the ventricle, and damming back of blood in the venous system—the former being the primary and principal difficulty—and the indications are to relieve the ventricles of work and give them strength, and at the same time to deplete the venous engorgement.

This last object must, indeed, be taken first.

Venesection, the most effectual means of relieving the right side of the heart, is very rarely applicable in dilatation. It is conceivable that under some pulmonary complications resulting from chill, the engorgement of the right ventricle and auricle might be such as to make bloodletting the less of two dangers; but, the initial fault being weakness of one or both ventricles, we cannot trust the heart to adjust itself to a rapid change of any kind, and the right ventricle may not be in a condition to take advantage of the relief afforded it. Usually, however, the venous engorgement is developed slowly, and the indications for venesection do not arise.

The application of six or eight leeches over the liver is safer, and it will usually effect all that can be done by direct abstraction of blood. The indication for this local bleeding is enlargement of the liver, and, when this is considerable, it rarely fails to afford striking relief. Very frequently, for example, the patient, who has previously been tortured by sleeplessness, will sleep at once and sometimes for a night or two afterwards. The reason for selecting the hepatic region for the application of the leeches is simply that pain and tenderness felt there are relieved; it is not supposed that blood is drawn from the liver, or that the same amount abstracted elsewhere would not be equally efficacious. The leeches will be followed by a hot poultice, which, besides encouraging the bleeding, will bring blood to the surface. When the liver is not enlarged, and especially if the right ventricle impulse and sounds are weak, there is no advantage to be gained by leeching.

Concurrently with the application of leeches a purgative will be given, which will deplete the portal system at the same time that the leeching depletes the systemic veins. Afterwards this will be the principal means of keeping down the venous engorgement. In a large majority of cases, indeed, we have to depend entirely upon purgatives for this purpose, as abstraction of blood by any method is inadmissible. It is not a matter of indifference what purgatives are employed. The object of a purgative is not simply to carry off as much fluid as possible and so drain the tissues. This may be the case, perhaps, in ascites from cirrhosis of the liver; but in heart disease of any kind, and especially in dilatation, much more is to be gained by rectifying the balance of the circulation, the kidneys will then often resume their function and remove the excess of liquid.

Purgatives, then, may be made to contribute to this; and

before considering them further we must refer to the first object of treatment, the relief of the heart from work and the increase of its vigour. Of these we can be much more sure of the first than of the second; we can more easily and certainly diminish the resistance in the arterioles and capillaries than we can lend strength and efficiency to the action of the heart, and, without removing the obstruction in the peripheral circulation, it might only do harm to incite the heart—weakened as its structures are—to greater effort to overcome it. Mercurial purgatives have this effect of diminishing arterio-capillary resistance and of lowering arterial tension, and therefore of relieving the heart. This is a fact of clinical experience and observation, and its explanation is a matter of secondary importance, but the hypothesis by which, as it seems to me, it is best explained, is that mercury influences the liver chemistry and promotes the elimination of impurities which, when retained in the blood, give rise to resistance in the capillaries. But, whatever the explanation, the fact that the arterial tension is notably lowered by mercurial aperients is one which is confirmed by daily experience. It is remarkable how frequently the statement recurs in works on heart disease that other remedies often fail to act until a dose of calomel or other mercurial preparation has been given.

Mercurial purgatives, then, have the double effect of depleting the portal system, which relieves the enlargement of the liver and the distension of the right side of the heart, and of diminishing the resistance in the peripheral circulation, and so relieving the left ventricle of stress. Very commonly the best of testimony as to the beneficial character of the result is given by refreshing sleep. The disadvantage, if such it be, that less fluid is carried off than by hydragogue cathartics is often compensated by an increased flow of urine; and elaterium, gamboge, pulv. jalap. co., and the like, when repeated, give rise to great exhaustion.

Calomel, then, or blue pill or grey powder, should be given in doses of from 1 to 5 grains, according to the urgency of the case, with colocynth and hyoscyamus or rhubarb, followed by some mild saline. After one or more full doses at the outset, a moderate dose may be given every second or third night. The acid tartrate of potash will often co-operate beneficially, both by its action on the bowels and on the kidneys.

The heart being relieved of work may be urged to more vigorous contraction by digitalis, strophanthus, spartein, squills, caffeine, convallaria, apocyanum, the special heart tonics, with which strychnine may usually be combined with advantage. In a case of extreme suffering, digitalis may be given with ammonia, ether, and nux vomica; in a more chronic stage, with iron, strychnine, and perhaps nitric or hydrochloric acid. Sometimes an effect can be obtained by giving citrate of caffeine in a pill, at the same time with digitalis in a mixture, when singly neither seems to be efficacious. Squills, again, may be given with digitalis, as in the well-known pill with mercury, or in some liquid combination. Next to digitalis stands strophanthus, which is a most valuable alternative when digitalis seems to produce sickness, as is sometimes the case, or when it fails to exercise a favourable influence on the heart. Sulphate of spartein I have seen to be of great service when digitalis and strophanthus appeared to have exhausted their influence. Of convallaria I have little to say. Apocyanum has, in one or two cases, seemed to carry off dropsy in a remarkable way, but one patient died suddenly when apparently just well.

It is not necessary to go into greater detail with regard to these remedies. Throughout a case of the kind the medical man has to fight, so to speak, with both hands, and continuous watchfulness will be necessary to meet the vicissitudes which occur, and many changes in method

DILATATION.

may be required, while the same principles are held in view.

The prognosis, as has been said, will be greatly influenced by the response to treatment. This will be energetic, especially in the matter of purgatives, in proportion as the symptoms are urgent, and if no favourable effect is produced, the prospect of recovery is very poor. Very commonly improvement takes place up to a certain point, and then progress seems to come to an end. This is a trying stage both to the patient and to the medical man. Change of remedies and new combinations must be tried, both in regard of the aperient and of the tonic, not frequently and capriciously, but with careful study of the results and due allowance of time for obtaining them. Sometimes it does good to suspend all medicines for a few days and start afresh.

REMOVAL OF DROPSICAL EFFUSIONS.

Under these circumstances, again, it may be of the greatest service to drain the fluid from the legs, even when the extent of the œdema is not such as actually to call for it. The good result of removing an ascitic accumulation, should this be present, may be still more striking. Even a moderate amount of effusion in the pleural cavity, such as we should not think of dealing with under ordinary circumstances, should be aspirated. A straw may turn the balance either way, and a very slight obstacle may prevent the heart from regaining control over the circulation. It is not always desirable to postpone the removal of fluid till the particular conjuncture described arrives; it may be an urgent necessity at a very early period. Usually, however, it is prudent to give remedies a chance before resorting to puncture or paracentesis. As regards the method of drainage to be employed, Southey's tubes are, in my opinion, much the best, whether for œdema or ascites, but particularly in case of ascites, when, of course, the tube is larger and longer than that used for œdema.

The feeding of the patient through the long course of treatment will be a task of extreme difficulty. We have to contend with nausea and distaste for food amounting to disgust; sometimes the sufferer positively cannot swallow anything requiring mastication. The object to be held in view is to keep down the volume of the blood while maintaining its quality. A small amount of solid or semi-solid food should be taken about every three hours. When the patient is not too ill to take his meals at the accustomed times it is a great encouragement to him to be allowed to do so, and he may then eat what he can of fish, fowl, tender meat, game, milk puddings. When the appetite is small the regular meals may be supplemented by intermediate nourishment, such as a beaten-up egg, a little milk, or perhaps a small cup of strong soup or beef-tea, or a little beef or chicken jelly, or meat extract. Soups and jellies have the disadvantage of containing little proteid and much liquid, and many extractives, but they are stimulants to the flagging heart. Potted meat sandwiches are a great resource, and the pulp of raw beefsteak can be given in this form, disguised by cooked meat or concentrated gravy. A German method of treatment is to feed patients suffering from cardiac dropsy entirely on raw ham.

The amount of fluid must be restricted as far as possible, especially that taken with food. Stimulants are usually imperatively necessary, but should be kept within limits known to and defined by the medical man. The patient is under a great temptation to resort to them for the relief of faintness, exhaustion, and nausea. Cream of tartar drink may be taken to quench thirst between meals, and in some cases a copious draught of hot water once or twice a day will run through the system rapidly and wash out the organs and tissues without augmenting permanently the volume of the blood or adding to the dropsy. When this is tried it must be ascertained definitely that the amount of urine is correspondingly increased.

A question which arises in almost every severe case is whether the patient must be urged to remain in bed or allowed to get up. Bed is undoubtedly the best place for him at first, during what may be called the crisis of the attack, for many reasons; the rest and warmth protect the heart from the strain of exertion and changes of temperature. On the other hand, the dyspnœa is usually worse in the recumbent posture, even with the shoulders raised, and may be intolerable unless the legs are allowed to hang down; not unfrequently it is simply impossible for the patient to remain in bed. A suitable chair, therefore, is always necessary, with support for the elbows, shoulders, and head, which can be taken advantage of in turn in the frequent changes of position to which the patient has recourse to ease his breathing or elude discomfort, and the quickness and ingenuity of the medical man or nurse in devising expedients may greatly alleviate his sufferings. A bed table or other form of support, upon which the patient may rest his arms or elbows and head when leaning forwards while sitting in his chair, will often be very useful. A patient will frequently sleep better in this position than in any other. Perhaps the most common state of things is that the patient is up during the day, and tries to spend more or less of the night in bed. When he cannot at once bear to go to bed at night, he may undress and sit in his chair wrapped in blankets near the bed for a time, when he will often, after a nap or two, be able to lie down and sleep.

Prognosis in the Earlier Stages.

We may now come to the prognosis and treatment of dilatation of the heart at a period when it has not given rise to the serious consequences just described, and an enormous degree of dilatation may be arrived at before such effects are developed.

As regards physical signs, the most important evidence will not be as to the actual size of the heart, but as to the strength remaining in its walls. A greatly dilated heart which is capable of giving a recognizable apex beat and fair impulse of right or left ventricle is more to be trusted than one which has apparently undergone less change in its dimensions, but the movements of which can scarcely be felt at all. There is always a possibility that degeneration may enter into the causation of the symptoms. It is better, again, that the sounds should be strong than weak, spaced than approximated, and it is important that the second sound of both ventricles should be well marked. Irregularity of action is, according to my experience, of less consequence than frequency, and I have learnt not to attach serious importance to the regular alternation of a strong and weak beat, so long as there is no great frequency with it.

The pulse is of importance chiefly as an indication of the vigour with which the blood is propelled into the arterial system. Dilatation of the heart often presupposes high arterial tension at the time of its production, and it is a favourable sign that a certain degree of this tension persists. When we find a short, small, low tension pulse, unless this is the result of treatment, it means that the heart has not sufficient strength to maintain the pressure in the arteries.

The probability of improvement or of prolonged immunity from ulterior consequences will next depend on the general condition of the patient, on the soundness and nutritional vigour of his tissues and organs, and especially of his blood-vessels; if he is weak and anæmic, on the cause and character of the impairment of health—whether it is inherent or accidental, whether, in effect, it is remediable or not. When the state of the heart and of the general health is attributable to alcoholic or other excesses, including the abuse of tobacco, to habits or mode of life or external conditions adverse to health, the future of the patient will depend on whether or not he is willing and able to renounce his self-indulgence, and relinquish his vices, or alter his ways and surroundings. Considerations of this kind are of the utmost importance, but they need not be enlarged upon here.

TREATMENT.

When we come to consider the treatment of cardiac dilatation in this stage, the first question to be asked is, What object may we venture to set before ourselves? What assurance can we give our patients? Can we ever reduce existing dilatation? Can we even arrest the tendency to increase, and avert or postpone the evil consequences? To no question can we return a more confident answer. Every day we see dilatation reduced in some degree in favourable cases, and, at the worst, much can be done to prevent the development of its ill effects. It has already been stated that in the acute aggravation of dilatation which gives rise to symptoms, the apex and extreme limit of dulness may be seen to return towards the normal position day by day, and the well-named " curable mitral regurgitation " of Dr. George Balfour is a form of dilatation incident to anæmia.

The means to be taken are in the first place the removal of any recognized cause in the habits, mode of life, and surroundings, so far as this is practicable. Next, to improve the general health by favourable hygienic influences and by suitable remedies. The amount of exercise to be taken will be a most important question, but no rule can be laid down applicable to all cases. Speaking generally, exercise is good, and whatever amount of walking can be done without breathlessness or exhaustion, and especially with enjoyment, that will be safe and beneficial, and, provided due prudence is exercised, walking up hill need not be forbidden. It is injurious as well as cruel to insist on an amount of exercise which distresses the patient. The exercise should be regular and daily, not spasmodic with intervals of inaction. A walk may be taken day by day with advantage which, taken only once a week, would be injurious. The exercise, again, should not come immediately after food, and it is better taken early in the day than in the afternoon or evening. A walk into the City, for instance, might do good, when a walk back after the day's work would do harm.

In some cases the treatment may have to be begun by actual rest in bed for two or three weeks, perhaps with massage, and the patient may have to resume exercise with just the same caution as food is resumed after hæmatemesis from gastric ulcer.

The question will now come before us whether we should recommend the Schott treatment or the systematic graduated walking up hill at high altitudes known as the "Oertel" treatment for heart disease. These have already been described and discussed, and they may both be of service in suitable cases. Besides the actual benefit that may be derived from either of these lines of treatment, we must take into consideration the fact that many people will submit to and carry out strict regulations, to which a certain amount of mysterious virtue is attached, who will not obey common-sense instructions. Fatty overloading of the heart in connection with obesity, and some cases of early dilatation will be benefited by the Oertel treatment; but the Schott treatment is more suitable and beneficial in cases of dilatation, more especially when the dilatation is due not to severe strain or over-exertion, but to an atonic condition of the cardiac muscle.

But it is not always possible either to modify the patient's life in such a way as to put an end to the conditions which tend to produce and aggravate dilatation of the heart, or to secure for him all the desirable hygienic influences. It would be useless, for example, to recommend a City clerk to relinquish the sedentary occupation which is his only possible livelihood, and little less vain to order a lady who has never walked in her life to take regular exercise. We can, however, do much to keep down the arterio-capillary resistance or high arterial tension, which plays such a fatal part in the production of dilatation of the heart; we can regulate the diet and warn against habits and occasions which are injurious, and, by securing efficient and continuous elimination from the liver and bowels, we can often make just that difference which is required for the safety and comfort of the patient. This can usually be effected by an aperient pill, containing a small dose of one or other of the appropriate mercurial preparations, taken twice a week, and we must not be afraid to order patients, whose hearts have given way under protracted arterial tension, to go on with such a pill for years. Tonics of various kinds may be given with advantage from time to time with acids or alkalies. It is not often that digitalis or remedies of the same class will be required continuously, though they may render service given from time to time.

CHAPTER XVIII.

STRUCTURAL DISEASE OF THE RIGHT VENTRICLE.

PHYSIOLOGICAL DILATATION—HYPERTROPHY AND DILATATION—DEGENERATION OF THE MUSCULAR WALLS.

DILATATION and hypertrophy of the right ventricle are perhaps the most common of all the structural changes to which the heart is subject. Some primary dilatation accompanies all severe exertion, and when the exertion has been inordinate it may be considerable and persist for some time, but unless the muscular fibres have undergone degeneration from age or abuse of alcohol, or a sedentary life, or from recent febrile disease, or diphtheria, there is a wonderful power of recovery. If over-exertion is habitual the dilatation will be neutralized by hypertrophy. In all cases of valvular disease of the left ventricle or of lung disease, such as bronchitis, emphysema, etc., which give rise to obstruction in the pulmonary circulation, dilatation and hypertrophy of the right ventricle follow as a necessary physiological result, as has already been explained in a previous chapter.

Dilatation and hypertrophy of the right ventricle are thus, for the most part, diseases in name only, and are really compensatory adjustments which neutralize more or less perfectly the effects of valvular or other disease of the left side of the heart, and when compensation is effectual become an indication of the degree of severity of the original disease.

PHYSICAL SIGNS.

Dilatation of the right ventricle carries the apex of the heart to the left and increases the horizontal dimensions of the heart. The apex beat may be found at or beyond the vertical nipple line, not unfrequently in the anterior or even mid-axillary line, when there is also enlargement of the left ventricle. The character of the apex beat depends more on the left ventricle than the right, but with much dilatation of the right ventricle it is usually diffuse. The area of cardiac dulness will be increased, and, on percussion, it may be found that the heart extends outwards beyond the point at which the apex beat was felt, and the line of dulness of the left margin of the heart will be more rounded and convex than normal. The area of dulness may extend to the right as far as or considerably beyond the right margin of the sternum in consequence of the dilatation of the right auricle, which usually coexists with the dilatation of the ventricle.

The degree of hypertrophy is estimated by the force of the right ventricle impulse as communicated to the hand applied over the lower left costal cartilages and the epigastrium, and also by the degree of accentuation of the pulmonic second sound. It is the degree of hypertrophy which is of vital importance, as on this depends the efficiency of compensation.

The **symptoms** usually associated with the changes in the right ventricle just described are pulmonary congestion, venous stasis, and dropsy, but these are not so much the effects of dilatation and hypertrophy of the right ventricle as the final results of the original disease of the left ventricle or lungs, which the right ventricle has been unable to combat efficiently.

T

It has seemed to me that our ideas of the effects of disease of the right ventricle have been too much based upon a study of the symptoms which attend affections of this ventricle secondary to disease of the left ventricle or its valves, which therefore are really attributable to the original disease.

Undoubtedly the supervention of dilatation of the right ventricle and of reflux through the tricuspid orifice allows back pressure to be brought to bear upon the veins, but this only intensifies pre-existing effects and symptoms, and makes no change in their character.

We are perhaps justified in assuming that the venous back pressure, to which insufficiency of the tricuspid valve will give rise, will in some degree produce the same results, whatever the state of the left side of the heart, and whether or not the tricuspid regurgitation has been caused by obstruction in the pulmonary circulation; but the conditions are fundamentally different when the tricuspid reflux is primary. It is not impossible, for example, that a sound and strong left ventricle may come to the aid of the right ventricle, just as the right ventricle so constantly comes to the aid of the left, notwithstanding the great length of the systemic as compared with the pulmonic circuit and the weak blood pressure in the systemic veins. The pressure which will cause the blood to spurt for two or three feet in venesection might carry a current through the capillaries of the lungs aided by the respiratory movements and the valves of the pulmonary artery.

It has, moreover, seemed to me that weakness of the right ventricle—the left ventricle being in a normal condition—has in some cases given rise to symptoms due rather to inadequate supply of blood to the left side of the heart than to damming back of blood in the veins. I have, for example, met with several instances of primary tricuspid regurgitation, either as a constant condition or coming on under very slight provocation. When any effect of this has been traceable it has not been breathlessness on exertion, but tendency to syncope. Perhaps this is what we ought to expect, since the occurrence of tricuspid regurgitation in the breathlessness of violent exertion has been regarded as a safety-valve action, while shortness of breath results from mitral regurgitation. A difference of symptoms ought to attend mitral and tricuspid insufficiency, one giving rise to turgescence and high pressure in the pulmonary circulation, the other to deficient supply of blood and low pressure. If mitral disease produces pulmonary symptoms, tricuspid disease may well produce systemic symptoms.

In a few cases which have come under my notice, in which the right ventricle has appeared to be predominantly or almost exclusively affected by asthenia or degeneration, the effects have been similar to those of tricuspid regurgitation.

A master in a public school, who had been accustomed to vigorous exercise, received a severe blow on the chest. He had for some time great pain in the cardiac region, and when he walked he soon felt faint. When I saw him some time after the injury he had had slight but distinct syncopal attacks. On examination, no valvular disease was present, and the heart was of normal size. There was a fair apex push in the normal situation, and the left ventricle first sound and the aortic second sound were normal in character. No right ventricle impulse, however, could be detected, and its first sound and the pulmonic second sound were very weak.

What had happened exactly cannot be stated, but from the contrast between the action and sounds of the two sides of the heart, it seemed as if the right ventricle had been in some way injured, and that its contractile energy was impaired.

The patient regained the power of taking exercise, and with this the sounds of the right ventricle became normal.

A patient, aged 72, who had never had a day's illness in his life, consulted me in 1879, complaining of failing vigour, giddiness on running to catch a bus, tendency to fall asleep in the day. He was constipated; the pulse, which was not more than 60, was sometimes tense, sometimes soft. The heart sounds generally were weak, the aortic second accentuated. In April, 1881, symptoms, which had previously been relieved, returned, and it was now found that over all parts of the right ventricle, and even over the pulmonic area, there was absolute silence. No impulse or apex beat could be detected, but at the apex the sounds were normal; the aortic second sound was accentuated. The pulse was 72, a little irregular, but fair in force and length. A month later he was much better. The pulse was 60, fair in strength and length; the left ventricle sounds were good, the right ventricle sounds faintly audible. From this time neither the right ventricle first nor the pulmonic second sound were ever at any time audible. The pulse varied considerably both in frequency and in tension, but it was usually well sustained, indicating considerable vigour of the left ventricle, and the left ventricle sounds were good. He had from time to time severe fainting attacks, which sometimes threatened to prove fatal. He was always worse when the bowels were not kept freely open. In June, 1883, the pulse is described as large, full, and tense; the aortic second sound was accentuated at the apex and in the right second space; there were no right ventricle sounds whatever. Towards the end of 1884, when not under my care, the bowels were allowed to get confined, and he fell into a condition of sopor, with incontinence of urine and fæces. He recovered from this condition after free purgation, and was again able to go about, though his mental faculties were impaired and he was childish; but in November, 1885, thrombosis of the left middle cerebral artery took place, giving rise to hemiplegia and aphasia, of which he died.

The entire absence of right ventricle sounds in this case was very remarkable, and I cannot doubt that there was really no action of this ventricle. As it seemed to me impossible that the pulmonary circulation could be maintained without its aid, I formed a very unfavourable prognosis; and when this was belied I watched the case with extreme care, first to make sure that my observation was not at fault, and next in order that I might arrive at some comprehension of the problem presented by the facts. The conclusion appeared to be unavoidable that the left ventricle was carrying on the circulation through the lungs; it was throughout capable of maintaining high tension in the arteries. The amount of blood passing through the pulmonary vessels under these conditions and reaching the left auricle would be easily influenced, and would vary greatly, and the fluctuating supply of blood to the left ventricle would account for the varying character of the pulse.

One of the most serious effects of weakness of the right ventricle is met with in disease of the mitral valve. When the mitral valve from thickening and shrinking of the curtains and tendinous cords become inefficient, the right ventricle is for a time the rampart by which the reflux of blood is arrested and a reinforcement to the crippled left ventricle. When we hear a systolic apex murmur telling of mitral regurgitation, the murmur itself gives no trustworthy information as to the amount of blood which is carried back into the auricle. We gather this mainly from the effects upon the right ventricle. The first of these is accentuation of the pulmonic second sound, indicating increased pressure in the pulmonary circulation, and following on this hypertrophy of the right ventricle, by means of which the obstruction to the passage of blood through the lungs is overcome. It is by the augmented strength of the right ventricle that the mitral leakage is neutralized and a working equilibrium established. The greater the regurgitation the greater the amount of hypertrophy required to compensate for it. The change in the right ventricle thus becomes, together with the accompanying dilatation of the left ventricle, a measure of the regurgitation.

When, therefore, we detect a mitral systolic murmur, we at once examine the right ventricle in order to gather from its condition information as to the amount of reflux which is not yielded by the murmur itself. This is specially the case when symptoms of failing compensation have set in. But if the right ventricle is in a state of degeneration or great weakness, these indications fail us altogether. The absence of dilatation and hypertrophy, instead of denoting comparatively slight regurgitation into the left auricle, shows that the right ventricle is incapable of coping with it. There is no right ventricle impulse, and the pulmonic second sound instead of being accentuated is weak, while dropsy and other evidences of serious stasis and back pressure in the venous system are prematurely developed for lack of hypertrophy or compensatory effect on the part of the right ventricle.

Under these circumstances the prognosis is extremely grave. The right ventricle is unable to come to the aid of the left, the mechanism of compensation makes default, and the back pressure bears at once upon the venous system. The fulcrum for some of our most efficacious therapeutic measures is missing. We dare not open a vein however great the respiratory embarrassment and cyanosis; the effect of leeching over the liver is less certainly good, and a dose of calomel is not well borne. Recovery is rare, and twice it has happened in my experience that during apparent convalescence, when an unfavourable prognosis seemed to have been belied, the patient has died suddenly when beginning to walk about, and the right ventricle has been found degenerated at the autopsy.

Similar conditions result from time to time from adhesion of the pericardium. The right ventricle suffers much more than the left from pericarditis. During the attack the muscular fibres immediately subjacent to the serous membrane, which are paralyzed by the inflammation, form an appreciable proportion of the thin wall of this ventricle, which is thus weakened and prone to dilate; whereas, although the superficial layer of the muscular fibres of the left ventricle is similarly paralyzed, they constitute a relatively unimportant part of the mass of the ventricular muscle. Again, the right ventricle is much more hampered by adhesion of the pericardium than the left, partly because its superficial area is relatively large, but chiefly because of the thinness of its walls; and when the adhesions are general, and especially if there is also adherence of the pericardium to the chest wall and diaphragm, efficient contraction of this ventricle must be impossible. Cases of this kind, then, are not uncommon. There is valvular disease, mitral or aortic. From the size of the heart, the position and character of the apex beat and impulse, the persistence of sounds in spite of the murmurs, there are grounds for concluding that the valvular lesion is not very great, but there is a premature development of symptoms. Under such circumstances, we may often confidently infer adhesion of the pericardium when it cannot be demonstrated by physical signs. The right ventricle may be obviously labouring, while the pulmonic second sound is less definite and pronounced, which will tend to corroborate the conclusion arrived at.

The right ventricle is undoubtedly sometimes the cause of sudden death, and when the heart is embarrassed or stopped by pressure upwards of the diaphragm by a distended stomach or colon, it must be on the right ventricle that the pressure takes effect. This part of the heart rests upon the diaphragm, and will be directly compressed when it is pushed up. Probably it is the diastole which is mostly interfered with, and it would seem that the proper expansion and filling of the ventricle must be impossible when the pressure upon it is such that the heart is carried up bodily by the diaphragm, especially when the ventricle is dilated and over-distended. A melancholy illustration of this occurred in my experience in the case of an eminent artist. He was suffering from mitral stenosis and regurgitation, and had overthrown the compensation established by hypertrophy of the right ventricle by serious imprudence in the form of over-exertion and exposure undertaken to remedy the effects of overwork. He was suffering in an extreme degree from distension of the right side of the heart, with tricuspid regurgitation, and especially from sleeplessness and dyspnœa, so that his misery was insupportable, and life was despaired of. The application of leeches over the liver, which was enormously swollen, and the administration of calomel, at once gave him sleep, and by a repetition of the leeches and regular employment of mercurial aperients, with the usual heart tonics, he so far recovered as to be able to leave his room, and his convalescence seemed to be assured.

One morning, after a hearty breakfast in bed, the nurse was about to wash his face and hands as usual, but he impatiently bade her give him the basin, and stand aside. He sat up in bed with the basin between his knees, and when the time came for washing his face, bent forwards over it. The pressure upwards of a full stomach caused by this movement brought the weak right ventricle to a standstill, and the patient fell back dead.

One cannot help being reminded, in relating this incident,

of the rough and ready, but effectual, way in which a man is brought to who faints after the severe exertion of a boat race. He is seated on the ground, and his body is bent forcibly forwards, so that his head almost comes to the ground between his knees, or, if he has fallen forwards over his oar, it is done while he is in the boat. The *modus operandi* of the remedy is pressure on the distended right heart, and, when emetics are resorted to in bronchitis, which has gone on to the production of cyanosis, the good effect is due, not only to the emptying of the bronchial tubes, but to unloading of the right auricle and ventricle by compression in the act of vomiting.

CHAPTER XIX.

FATTY DEGENERATION.

DISTINCTION BETWEEN FATTY INFILTRATION OF OBESITY AND FATTY DEGENERATION—CAUSATION OF FATTY DEGENERATION—SYMPTOMS—PHYSICAL SIGNS—DIAGNOSIS—PROGNOSIS—TREATMENT.

No form of heart disease is regarded with so much apprehension as fatty degeneration. More than any other, it carries with it the danger of sudden death and the liability to angina pectoris, and, although happily it is not very common, it would be a most important acquisition to be able to make the diagnosis with certainty at an early period.

It must be understood from the first that the fat-laden heart of obesity does not come under the designation of fatty degeneration, and it may be dismissed from further consideration with a few words. In advanced life there is a tendency to the formation of adipose tissue beneath the visceral pericardium, especially along the course of the coronary arteries. In obese persons the amount may become considerable, so that the entire heart may be encased in fat, and adipose deposit may penetrate between the muscular fibres. When such is the case, the heart will be hampered in its action, and a further source of embarrassment will be present in fatty deposit on the surface of the diaphragm and in the omentum.

In fatty infiltration there is a deposit of fat between

the muscle fibres, which may be traced, on careful examination, as a down-growth from the overlying layer of fat around the heart. There is no substitution of fatty material for muscular substance, as is the case in fatty degeneration. It is possible that compression of muscle fibres by intervening fat may lead to atrophy of some of them, and it is of course possible that there may be fatty degeneration as well as fatty infiltration, but the two are quite distinct processes, and are due to different causes.

Extreme and even distressing shortness of breath may be produced by such fatty deposit and infiltration, and not uncommonly there is a certain degree of œdema about the ankles and along the tibia at night, especially in hot or relaxing weather. The condition, however, is not attended with the same danger as actual degeneration of the muscular fibres or with the characteristic symptoms.

In fatty degeneration the heart substance will be pale and softer than normal, so that the finger can be readily thrust into it; the musculi papillares will usually have a streaky appearance, the so-called "tabby-cat" striation, due to pale strands of fattily degenerated muscular substance, being interspersed among healthy fibres.

Microscopically, on staining with osmic acid, it will be seen that an actual substitution of fat for muscular tissue has taken place, the deposits of fat being in the muscle fibres themselves, and not between them. The tiny globules of fat first make their appearance at the poles of the muscle nuclei, but eventually the degenerative process extends throughout the muscle fibre, so that it loses its striated appearance, and is seen to be filled with globules of fat.

When unstained, the granular appearance of the degenerated fibres might be mistaken for the condition of "cloudy swelling" or albuminoid degeneration; but the distinction is readily made by treating the section with acetic acid, which does not affect fat globules, but dissolves the albuminoid granules, or by staining with osmic acid, which gives quite characteristic appearances.

Causes of Fatty Degeneration.

Its causation, which must be considered, as it bears upon the prognosis, is in some cases clear, that is, when there is disease of the coronary arteries or obstruction of these vessels by any other means. The heart—perpetually at work—cannot afford to be mulcted of its full supply of blood. When, from any cause, this is defective, the wear and tear of the muscular fibres, which must go on, is not repaired, and their structure breaks down. Whether the *débris* actually forms the fatty granules and globules which are found within the sarcolemma, or the fatty particles are substituted for the atrophied sarcous elements, is not, perhaps, a settled question. The important point is that the primary change is atrophy of the muscle substance, the invasion of the fibres by fatty matter being secondary to this and consequent upon it.

Disease of the coronary arteries, being thus a cause of fatty degeneration of the heart, the existence of conditions which may lead to the implication of the coronary arteries or their orifices in morbid processes, will warrant a suspicion that any cardiac weakness which may be recognized is the result of degeneration. For example, an aortic murmur coming on after middle age may not indicate serious valvular lesion, but, as it is probably the result of atheromatous changes in the valves or arterial walls in close proximity to the orifices of the coronary arteries, there is reason to apprehend that the disease may cause obstruction here or may have extended to the vessels themselves, and progressive weakness of the heart, were this to supervene, would be attributable to degenerative change in its walls.

A like apprehension attaches to syphilitic disease of the aorta and its valves, which is not very uncommon in early middle life. Acute aortitis, again, is recognized mainly by a train of effects on the heart, produced in the same way by blocking of the mouths of the coronary arteries.

But there may be fatty degeneration of the heart when the coronary arteries are healthy. It is usually present—sometimes in a very advanced degree—in pernicious anæmia, and granular degeneration, which is sometimes considered to be an acute form of the disease, is a constant effect of severe typhoid fever and of fatal phosphorus or arsenical poisoning. Cases occur from time to time in which a patient convalescing from typhoid dies suddenly on sitting up in bed. Here, again, in fatal anæmia and enteric fever the process must be the resultant of a balance on the wrong side, as between the catabolic and anabolic operations, disintegration and repair, but it is now the quality of the blood which is at fault, not the supply, and in typhoid fever there is also the injurious effect on the nutritional processes of long-continued high temperature. It is not to be wondered at that in pernicious anæmia and fever the heart suffers more than the voluntary muscles, since these are at rest, and there is no functional wear and tear, whereas in the heart this is continuous and excessive. From what takes place in typhoid fever again, it is seen how rapid degeneration may be.

Diabetes, alcoholic excess—tippling rather than drunkenness—a sedentary mode of life, may conduce to fatty degeneration of the heart, probably through deterioration of the blood, or degeneration may be secondary to myocarditis. Cases are met with for which no explanation can be found, and we are almost compelled to assume that there may be a defective assimilative action in the muscle cells of the heart, or possibly some unrecognized blood condition.

Symptoms.

There is little that is characteristic in the symptoms unless we consider angina pectoris to be such, and definitely associate it with fatty degeneration of the heart. The relation between the two is undoubtedly very frequent but is not constant, and angina is therefore reserved for special and separate consideration.

In a large proportion of cases the subject of this affection has had no ailment which has led him to consult a medical man when he is overtaken by sudden death during exertion or excitement, or after a full meal. Or, the excitement and exertion may be passed through safely and death follow some hours later, next day even. Among the causes which precipitate a sudden fatal termination, dilatation of the stomach is frequent. Digestion is usually imperfect from advancing years, or as a result of sluggish circulation due to the state of the heart, and the tone of the muscular coats of the stomach is impaired, allowing of passive distension by the contained gases, the products of fermentation. The upward expansion of the stomach is moreover often facilitated by a weak and relaxed condition of the diaphragm, so that the upper line of gastric resonance can not unfrequently be traced horizontally across from the root of the ensiform cartilage to the usual situation of the apex beat in the fifth space. Such a condition is attended with immediate danger and may cause sudden death by pressure on the heart, long before this would have resulted from the state of the heart alone.

Rupture of the heart is one mode of termination, and this may take place on very slight provocation. Sometimes the patient has been engaged in his usual avocation up to the moment of its occurrence. In one case which came under my observation, an old gentleman of quite retired habits, with nothing beyond the weakness incident to age, was heard to knock at the wall against which his bed was placed, and was found dead, the bedclothes scarcely being disturbed. A neat slit was found in the left ventricle near the apex close to and parallel with the septum.

The bearing of such occurrences on prognosis is direct and simple. No doubt in many cases of sudden death there have been warnings which the patient has ignored or has not spoken of. These will sometimes be acknowledged in the course of examination when they have not been mentioned spontaneously.

When the course of the disease has been sufficiently chronic to permit of the recognition of symptoms, which in my experience is chiefly when the degeneration is secondary to change in the coronary arteries or to old-standing hypertrophy, with or without dilatation, they will be such as are produced by a slackening circulation, and they are not so different from those attending dilatation as to permit of any distinction being drawn between the two conditions in an early stage, without physical examination. There may, perhaps, be greater fluctuations in dilatation, though even in degeneration there may be great temporary improvement under care and treatment. In advanced stages characteristic differences make their appearance. The symptoms of advanced dilatation have already been described; those attending degeneration are evidences of heart failure of another kind. A noteworthy point is that well-marked dropsy is rare, and probably never occurs in uncomplicated degeneration. The significance of this is that the special effect of the disease is defective pressure in the arterial system; and it is to this are due the syncopal, apoplectic, and epileptiform attacks, which, together with the angina pectoris, are the most characteristic later effects of fatty degeneration.

The syncopal attacks vary greatly in intensity. So far

as they have come under my observation, they have been marked rather by duration than intensity, and have rarely been so complete as to be attended with absence of consciousness; they have usually been accompanied by prolonged coldness of the extremities and of the surface. I have not met with instances of sudden and complete loss of consciousness and immediate recovery as in dilatation. These syncopal attacks are very significant, and are often premonitory of fatal syncope.

The apoplectiform seizures are very remarkable, and in the absence of history and without examination they are not distinguishable from the apoplectic condition resulting from cerebral hæmorrhage. The patient is unconscious; the respiration, if he is allowed to lie flat on his back, may be stertorous—though stertor, after the teaching of Dr. Bowles, ought to be eliminated from the symptomatology of apoplexy—and there may be hemiplegia, though this will be fugitive. Cheyne-Stokes breathing, which was first observed in connection with fatty degeneration of the heart, has not been present in the few cases which I have actually seen in the apoplectiform state, while I have met with it in a very large number of cases of uræmic coma and in connection with serious consequences of high arterial tension. On examination of the pulse and heart, however, it will be clear that there cannot have been sufficient pressure in the arteries to rupture even the most degenerate vessel, and, on the other hand, thrombosis or embolism is not competent to produce unconsciousness of the character and duration of these attacks. According to my experience, the patient is never quite the same after an apoplectiform attack; he is feebler in mind and body, and sometimes increasingly liable to syncopal attacks.

The epileptiform attacks are not often violent, but resemble *petit mal* rather than a typical epileptic fit; while, however, the convulsion may not be so severe, there is profound unconsciousness—not like epileptic coma, but of a syncopal character—and the pulse may be extremely infrequent, sometimes less than twenty in the minute. In my judgment, the heart failure manifested by the slow pulse and the consequent arrest of the cerebral circulation are the cause of the fits, and it is not the epileptiform attack that affects the action of the heart.

By the time any of these forms of attack occur the diagnosis of fatty heart is usually sufficiently clear; but I have had under observation a case of very slow pulse with *petit mal*, in which the strength and volume of the pulse and the degree of impulse of which the heart was capable precluded the idea of advanced degeneration.

An important question is whether there is anything characteristic in the appearance of a patient suffering from fatty degeneration of the heart? A greasy state of the skin with a sallow pallor of the face has been described, and if such a condition has supervened upon a previously healthy complexion the change would have significance, but nothing of the kind is present in a large majority of the cases. Many of the subjects of the disease retain the look of health for a long time, and even up to the moment when the heart ceases to beat. The degeneration may be due to a local cause—obstruction of the coronary arteries; and, even if a tendency to general deterioration of the tissues is present, the change mostly advances so much more rapidly in the heart than elsewhere, that there is no time for it to become conspicuous in the skin. The picture appears to have been drawn from cases of a universal chronic degeneration of vessels and heart. The arcus senilis again, which has been said to indicate the existence of cardio-vascular degeneration, has no such significence.

Physical Signs.

The most constant and significant feature of the pulse is that it is short and unsustained. The size of the artery at the wrist and the condition of its walls may vary greatly. When the arterial coats are healthy they are apt to feel extremely thin. The pulse rate may be regular and about normal, or extremely irregular both in force and time, and it may be frequent or slow. A very slow pulse with extreme low tension is most characteristic, but then it is the most rare.

The physical signs may be described as negative. Unless degeneration has attacked a heart already enlarged the size will be normal. If the fatty change is at all advanced, impulse can neither be seen nor felt, or, if perceptible, it is only as a faint vibration. A heart in this condition is incapable either of giving a distinct push or of maintaining continuous pressure in the arteries. The sounds are weak, sometimes so weak as to be almost inaudible; but except that the first is short, there is nothing abnormal about them; the intervals, again, are usually normal. The very absence of physical signs, such as murmur, or conspicuous modification of the sounds or intervals, or disturbance of the relation between the two sides of the heart, or increase of dimensions, when symptoms of serious slackening of the circulation are present, and especially when there have been anginal, or syncopal, or apoplectic attacks, adds gravity to the case.

But a weak, short, unsustained pulse is common as a constitutional peculiarity, or may at any period of life be simply a result of general debility, and impulse and apex beat may be entirely absent, and the sounds may be short and weak. In young people there is no danger of such weakness being taken to be indicative of degeneration of the heart, but it may arouse anxiety after middle age, especially if there is also irregularity in its action.

It is important to be able to distinguish between functional weakness of this kind and weakness arising from organic disease. Usually this is accomplished by making the patient walk briskly. A few steps will often be sufficient. If the heart is sound it rises to the occasion. The pulse, and beat, and sounds are all more distinct, and strong, and regular, whereas the fatty heart "goes to pieces," and the pulse becomes irregular and shorter than ever, or may even disappear.

Until the disease is far advanced the diagnosis of fatty degeneration of the heart is not easy, and is scarcely to be made without more than one opportunity of examination. When the diagnosis has once been made, the prognosis, for the most part, can contemplate only one result; a fatal termination is merely a question of time and circumstance. Excluding cases in which death has been sudden without warning, the shortest period in my experience over which characteristic symptoms have extended, together with recognized physical signs, has been about six weeks; several patients have survived the diagnosis two years before justifying it by dying suddenly. But circumstance as well as time enters into the question; a slight effort, or a fall, a little hurry or excitement, too hearty a meal, an attack of flatulent indigestion or constipation, a chill, may hurry on the fatal termination; and on the other hand, judicious care may postpone it till the heart is completely worn out and comes to a standstill.

The question must be asked, Is fatty degeneration of the heart ever cured or arrested? If the granular disintegration which is produced by typhoid fever is to be included under the term, the answer must undoubtedly be Yes. The heart may ultimately regain structural soundness and functional vigour when during the fever the first sound has been completely lost and the impulse has been scarcely perceptible; and when degeneration has been the result of other forms of blood poisoning or deterioration, it ought to be possible, and now and then to occur, that recovery of the heart should follow a return to a healthy state of the blood.

More than ten years ago I came to the conclusion that a gentleman still living, aged at that time about 55, was suffering from fatty degeneration of the heart. Spare in habit, strictly moderate in eating and drinking, regular in taking exercise, and a great pedestrian, he rapidly lost strength without recognizable cause, became breathless on very slight exertion, so that he could scarcely walk slowly a hundred yards without actually stopping, either to get his breath, or on account of anginoid pain. On one occasion at least, while sitting in his chair he became suddenly pale and unconscious, his head fell on his chest, and the jaw dropped. With this change in his health, the pulse and heart were extremely weak. He would never relinquish exercise, but continued to walk, however slowly and at whatever cost of pain and distress, every day, exercising great self-command and measuring his strength very exactly. Little by little he gained ground, and he is now in fair health, but capable of very little in the way of work. It should be added that never at any time were his intellectual faculties at all affected.*

This case may have been an instance of arrest and partial recovery.

TREATMENT.

It must be acknowleged at the outset that it is not in our power to modify in the least degree the condition of the cardiac muscular fibre when far advanced in fatty degeneration, and that we can do very little, if anything, to arrest the

* Since the above was written the patient has died suddenly in bed.

progress of the deterioration when it has reached a stage at which it is recognizable either through symptoms or by physical signs. Even at a very early period we should doubt the possibility of reversing a process which in some cases is an inherited tendency to natural decay at a given time of life, in others an effect of imperfect blood supply due to narrowing of the coronary arteries.

If, indeed, the degeneration has been the result of acute disease, such as typhoid fever, time and care will bring about a restoration of the muscular fibres; but this is not fatty degeneration, in the true sense of the word, but is rather a secondary consequence of the condition of so-called cloudy swelling, which is a result of prolonged fever.

When, again, the condition is not true degeneration of the muscular fibres, but a fatty heart, due to a sedentary life, with privation of fresh air and neglect of exercise, together with undue indulgence in alcoholic drinks and in the pleasures of the table, there is not then, at any rate till an advanced stage, true fatty degeneration, but a deposit of fat between the muscular fibres or fatty infiltration. In such cases careful dieting and graduated exercise, such as the Schott or the Œrtel treatment, may reverse the degenerative tendency and even cause a gradual absorption of the fatty tissue already deposited in and around the heart as well as elsewhere.

While, however, acknowledging the limitation of therapeutics in dealing with the organic change, we are not altogether powerless to avert its consequences, and by so doing to prolong life. We see from time to time, on post-mortem examination, the heart so far gone in fatty change that it is scarcely recognizable as muscle, either to the naked eye or under the microscope. Such change must have been long in progress, and the subject of it must have lived for months if not for years, while the slightest obstruction in either pulmonary or systemic circulation would have brought the heart to a standstill, and a shock, a fall, or an indigestible meal, would have been fatal. We meet with cases of this kind during life, in which the decay of the mental and bodily powers is so slow as to be almost imperceptible, or in which thrombosis of one cerebral vessel after another brings the patient to a state of dementia or bed-ridden paralysis from general or local cerebral softening. Or the immediate cause of death may be senile gangrene. It is not that results such as those just named are desirable—death would be preferable were we allowed to choose—but cases of the kind serve, with others, to show that prolongation of life is possible when the central organ of the circulation can barely keep the blood in motion, and to illustrate the conditions under which this is observed. These conditions are a gradual diminution of mental and bodily activity, together with attention to diet and regulation of the bowels. The setting in of softening of the brain, or an attack of paralysis, not unfrequently seems to put an end to cardiac symptoms and to prolong life; the sufferer is no longer his own master; he cannot undertake business or go about, his food is under orders, and the action of the bowels is known to others besides himself.

If, in an early stage of fatty degeneration of the heart, the same command over the patient's mode of life and the same knowledge of the state of his secretions were attainable, not only might life be made longer by many years, but much suffering which is seen to arise out of this form of disease might be averted. It is unnecessary, and it would be impossible to enter into particulars with regard to the amount of work and exercise to be permitted, or the quantity and kind of food to be allowed; the latter may, and indeed in most cases must, be liberal and varied, but precautions must be taken against an inordinate appetite, and it is always safest to let some judicious relative or attendant who knows his likings and what suits him, help the patient at meals, and decide for him what dishes and what quantity will be good for him, acting under the advice of the medical man.

Proper regulation of the bowels is of the utmost importance. As years increase people are apt to become less observant, and to take it for granted that, so long as the habitual regularity in going to stool obtains, the action of the bowels is satisfactory and efficient, whereas it may be that the evacuation is much too small in amount. Accumulation thus gradually takes place, and it is not uncommon for a second daily call to relieve the bowels, resulting from this, to be regarded as evidence of improvement in their action. The statements of patients, then, with regard to this function, are not always trustworthy, and the testimony of a competent observer, or inspection by the medical man, is necessary. The quantity of fæcal matters which may unconsciously accumulate in the colon is astonishing, and the prevention of such an occurrence is essential to the well-doing of a patient whose heart is organically weak. Palpitation and oppression are the smallest of the evil consequences which follow; fatal syncope may be induced, or such weakness of the heart's action as may be attended by complete prostration of strength which, once setting in, may last for weeks or months; or there may be cerebral symptoms, complete loss of memory, with childishness and torpor. For the regulation of the bowels mild aloetic aperients are best, with pil hydrarg. and colocynth occasionally in small doses, if there is arterial tension. Any tendency to flatulent distension of the stomach must be counteracted, as far as possible, by careful dieting and the administration of alkalies and carminatives. Bitter tonics may be given, and massage or gentle exercise is often of great service. Extremes of heat and cold should be avoided, and a dry, bracing, healthy spot in the country should be selected as a residence.

CHAPTER XX.

ANGINA PECTORIS.

CHARACTERISTICS OF TRUE ANGINOID PAIN—DURATION OF ATTACK—ASPECT OF PATIENT—EXCITING CAUSES OF PAROXYSM—PATHOLOGY AND ÆTIOLOGY OF TRUE ANGINA—THEORIES AS TO CAUSE OF THE PAIN—PROGNOSIS—TREATMENT.

WHILE heart disease generally, of whatever kind, is remarkable for the almost entire freedom from pain—so that, when patients come complaining of pain in the cardiac region it is a presumption against the existence of any serious organic affection of the heart rather than an indication of any such change—there is one form of pain in and around the heart, angina pectoris, which is very definite and constant in its significance of disease and danger.

In a characteristic attack of angina, there is intense pain in some part of the cardiac region—in the left breast, or behind the sternum, or across the chest, at its upper part usually, but occasionally lower down, with radiation down the left arm. Accompanying the pain is a sense of utter powerlessness and extreme fear and dread. The patient stands still, not daring to move or breathe, and feels as if he were in the act of dying. He will say afterwards that if the pain had lasted another moment he must have died. In no other condition is the physical agony of dying realized in anything like the same degree. The two elements of pain and sense of dying coexist in a true paroxysm of angina, and are almost equally characteristic.

The pain differs in character and situation and in intensity in different cases. Some sufferers will say it is indescribable—nothing in their previous experience suggests even a comparison; others speak of the pain as severe cramp in the heart, or as if the heart were gripped by an iron claw; while pain of a shooting, neuralgic character, sometimes intermittent, sometimes persistent, seems to radiate from the chest to the left shoulder, the inner side of the arm, the forearm, and the ring and little fingers. Occasionally there is a sensation as of the wrist being grasped so tightly as to cause pain. With the pain in the heart there may be pain down both arms or shooting up into the left side of the neck, very rarely in the right arm only. Occasionally the pain may be felt first in the wrist or arm and seem to travel up to the chest, or may come in the inner side of the arm as a kind of warning of an attack. Another description of the pain is that it feels as if the sternum were being crushed back to the spine, or, again, as if the whole chest were being held in a vice. In other cases the pain is compared to a bar of iron across the upper part of the chest; in others, again, to a ton weight upon the lower part of the chest. The ramifications of the cardiac plexus and its communications with other nerves make the radiation of pain in all the various directions enumerated comprehensible, and the nerve of Wrisberg has been specially instanced as explaining the pain in the left arm, but no explanation can be given why in one case the pain is felt in one part of the cardiac region, and has some particular character, and takes a given direction down one arm or both or through to the back, while in another case the seat, character, and extension of the pain are quite different. It is not a pressure effect on the plexus outside the heart, neither heart nor aorta being necessarily enlarged, and extreme fusiform dilatation of the arch of the aorta being common without confirmed pain; and there can be no stretching or mechanical irritation of the ramifications beneath the endocardium at all comparable to that which takes place in acute dilatation of the heart. It seems to me probable that the pain is really central, and that the radiation of irritation giving rise to its extension takes place in the spinal cord.

An interesting point is that at the end of a paroxysm there is usually flatulent eructation from the stomach. The attacks are therefore very commonly attributed to flatulence, and distension of the stomach by food or gases may undoubtedly be, and often actually is, an exciting cause, but more frequently the sensation as of wind on the stomach is only a part of the general commotion, and is due to communicated or sympathetic irritation of the gastric distribution of the vagus, the cardiac branches of which are primarily implicated. The escape of gas from the stomach is often a signal that the paroxysm is over rather than the means of bringing it to an end. Occasionally there is a vehement necessity to pass urine, although the bladder may at the time be empty.

The duration of the attacks is very varied; sometimes it can be reckoned in seconds. Most frequently, perhaps, a paroxysm will last a few minutes, but I have known a patient sit in the same position almost through an entire night, not venturing to make the slightest movement and scarcely seeming to breathe, while the perspiration rolled off his forehead and came through his clothes. According to my experience, it is when the attack comes on in the night, without provocation by exertion or exposure, that it is protracted. When it is started by exertion it generally ceases soon after the exertion is left off.

While it would not be justifiable to say that a patient was the subject of angina pectoris unless he had had one or more paroxysms of intense radiating pain, associated with a sense of immediately impending death, it must be admitted that attacks of true angina occurs which fall short of the typical development. For example, when a patient has been taught prudence by one or more bad attacks, he may, by standing still on the first warning, or by taking remedies, cut short the paroxysm, which will then have been represented only by the initial pain in the breast or arm without the mortal dread. It is possible, therefore, that before any characteristic attack has occurred, pains of a similar kind and intensity, disregarded by the patient or relieved by rubbing the chest or arm, may have the same significance as a fully developed paroxysm.

Again, a patient who has had attacks of true angina may cease to suffer pain, but may have attacks of what he calls faintness, in one of which he ultimately dies. These, which have lost their title to the name angina, have an equally serious significance. They are sometimes called angina pectoris sine dolore.

The aspect of the patient is one of extreme anxiety or alarm. He is usually pale and often livid round the mouth, but it is said that sometimes the colour does not change. A cold perspiration usually bursts out on the forehead, and may be so copious as to drip off the face. The pulse, in the rare instances in which I have had the opportunity of examining it during a paroxysm, has been irregular, small, and weak. In some cases it has been reported to be very small from contraction or spasm of the arteries. In others, again, it has scarcely been affected at all.

Great importance attaches to the exciting cause of the paroxysms. In the first instance they are almost always brought on by exertion. The patient, while walking perhaps more sharply than usual, or uphill, or against a wind, is more or less suddenly arrested by pain in the chest, with a feeling as if the heart were about to stop and he to fall down dead. On standing still the pain gradually passes off, and he is able to resume his walk, but only feebly and gently. For a while the attacks only occur when provoked by exertion, but more and more easily as time goes on, and they tend to become more severe. They are more readily induced when a walk is taken, or any imprudent exertion, such as stooping, drawing on boots, pulling open a drawer, pushing up a window, is made soon after a meal, especially after breakfast. External cold, again, predisposes to an attack, and exercise, which can be taken with impunity in mild weather, brings on a paroxysm if the air is cold and damp. Attacks, again, may be brought on by indigestion or constipation, apparently either through reflex disturbance of the heart, or as a result of pressure from the distended stomach or colon carrying the diaphragm upwards and obstructing mechanically the action of the heart and the expansion of the lungs.

They are also liable to occur during the night, and may be induced in various ways. The contact of cold sheets may have this effect by causing contraction of the peripheral arterioles, and thus throwing increased work on the heart; or the upward pressure of the abdominal viscera, on assuming the horizontal position, may embarrass the heart. Not unfrequently an attack comes on after sleep, when the vigour of the circulation has run down; when probably also there has been evolution of gases in the stomach and intestine, and distension of this viscus or of the colon giving rise to pressure on the diaphragm.

It is clear that the great exciting cause is a demand for increased effort on the part of the heart to which it is not equal, or, what is equivalent to this, interference with the movements of the heart by a dilated stomach and colon.

CAUSATION OF ANGINA.

The conditions of the heart associated with angina pectoris are varied, but perhaps the most remarkable and significant point in the relations between heart disease and angina is that angina does not attend the chain of events through which stenosis or incompetence of the mitral valve proves fatal, and is not among the symptoms which arise out of the valve lesion and its effects upon the heart. This fact was duly emphasized by Dr. Walshe, in his classical work on the heart, and no exception to it has occurred in my experience. I have, indeed, known instances in which, after attacks of angina have occurred at intervals for many months, mitral regurgitation has supervened with dilatation of the left ventricle, and concurrently with the establishment of so-called mitral symptoms—pressure in the pulmonary circulation, dilatation of the right side of the heart, and dropsy—the angina has ceased. In these particular circumstances Dr. George Balfour's view, that the giving way of the mitral valve may be an advantage to the sufferer from aortic disease, is perhaps justified.

Aortic stenosis may be attended with true angina, as may also aortic incompetence and a combination of the two conditions of the aortic valve. In association with aortic valvular disease angina may be met with in early adult life, and may continue for many years without proving fatal. The sense of impending death is, however, not fully pronounced in many aortic cases.

Adherent pericardium appears in some cases to be a factor in the liability to anginoid attacks when it co-exists with aortic valvular disease, but in my experience it has not given rise to angina when no other lesion was present.

Injury to the root of the aorta has been known to give rise to angina. I have had a case under observation for several years, in which a severe crush of the chest gave rise to a double aortic murmur and to distressing attacks of angina. For a time the attacks came on very frequently, even while the patient was kept in bed, and they continue to occur on very slight provocation, requiring frequent recourse to nitro-glycerine, which the patient takes in considerable quantity. There has been scarcely any compensatory hypertrophy and dilatation in this patient, and he has never been able to work.

In aortitis there is usually angina, the attacks at first slight, increasing in intensity and duration, and coming on more frequently as the disease advances. The heart rapidly becomes weaker without notable enlargement, the impulse more feeble, the sounds weak and short. Both the angina and the weakness of the heart point to interference with the coronary circulation, and the orifices of the coronary arteries are found small and contracted by the swelling of the walls of the aorta.

A perfectly characteristic attack of angina has been described to me as having occurred in intermittent fever, and serious weakness of the heart was left behind for some time. Angina, again, has sometimes been an incident of diabetes, possibly from high arterial tension, which is commonly present in this disease late in life. Occasionally, however, a series of severe anginoid attacks, occurring at short intervals, has been followed by rapid heart failure and dropsy, suggesting that the angina was symptomatic of myocarditis.

Attacks of pain in the region of the heart of various kinds, some being true angina, are spoken of as gouty, sometimes, no doubt, in order to disguise the real nature of the paroxysms from a nervous patient to whom the knowledge might be dangerous or fatal.

In a very large proportion of the cases in which angina has proved fatal, the heart has been found, when examined after death, to be in a more or less advanced stage of fatty degeneration, and in most of these again there has been disease of the coronary arteries, very commonly so far advanced as to have reached the stage of ossification or calcification. Sometimes these vessels can be dissected out from the auricular grooves as rigid calcareous tubes. The fatty change in the walls of the heart may be so far advanced that the fingers sink into its substance on very slight pressure, and that scarcely a trace of muscular fibres can be found on microscopic examination. On the other hand, the degeneration may be comparatively slight, being evident to the naked eye only as yellow striæ or patches in the ventricular walls and in the papillary muscles. The microscope, however, will show fat granules in those parts of the heart which to the eye and touch seem normal, as well as advanced fatty change when degeneration has given rise to yellow striæ.

In some cases the morbid condition found is fibrosis, general or local, apparently from myocarditis. Sometimes a distinct history of an attack of myocarditis is obtainable by questioning the patient. There may, however, be little or no recognizable change in the walls of the heart, especially when the first attack has proved fatal, or death has supervened after only a few paroxysms. It is probable, however, that in such cases something will be discovered, on minute examination, perhaps the obliteration of a branch of a coronary artery by endarteritis, or its obstruction by an embolus or thrombus. Something certainly must have happened.

If we now try to bring to a focus the more important conclusions regarding angina pectoris:—First as to the condition of the heart during the attacks. This has been generally supposed to be one of spasm, but there are great difficulties in accepting this view, and probably ideas as to what is meant by spasm of the heart in the anginal paroxysm by those who have employed the term have been diverse and very often vague. If by spasm of the heart is understood tonic contraction or an unrelaxing systole, this is certainly not the condition present. The heart has never been found in this state after death, and in most cases is absolutely incapable of such contraction from the state of its walls. No pulse would be possible were the heart in a spasm of this kind, and the pulse, though small and often irregular, can usually be felt. It has, indeed, in some cases been apparently unaffected by the paroxysm.

But by spasm may be meant an irregular and partial contraction like cramp in voluntary muscles, or a fibrillar contraction, such as is sometimes induced by faradic currents in muscle under experiment. The late Dr. Matthews Duncan, in the last conversation I had the honour to hold with him, suggested that the state of the heart in angina pectoris might be like hour-glass contraction of the uterus. He had probably at that time experienced the pain. Views of this kind cannot be proved to be wrong, but objections might be raised, and, for my part, I have to admit that I have no clear and definite idea of the state of the heart during the paroxysm.

The central fact and essential significance of angina is that stress is put upon the heart, to which, for the moment, it is unequal.

One of the main causes of such stress is persistent resistance in the peripheral circulation, or, in other words, habitual high arterial tension, and we owe to Dr. Lauder Brunton the knowledge that in many attacks of angina there is an aggravation of habitual high tension by a general contraction of the arterioles. But the habitual state of the arterial circulation may possibly be one of relaxed arterioles and capillaries and low tension, so that the heart has no abnormal resistance to overcome. Here sudden general arterial spasm would put the heart to greater stress than if the habitual tension were high, since the contrast between the work demanded would be greater.

Angina Vasomotoria.

When the paroxysms of angina can be distinctly traced to arterio-capillary resistance, or, when in the case of a patient subject to angina the usual condition of the circulation is one of high tension, the term "angina vasomotoria" may, perhaps, be appropriately employed. It is easy to imagine that stretching of the muscular fibres of the heart in the endeavour to overcome the resistance in the arteries might cause pain. But even when the vasomotor element is most potent, another factor must enter into the causation. Nothing is more common than high arterial tension, and it is met with in an extreme degree and produces fatal results without angina by ruining heart or arteries, or both, in hundreds of cases for one in which angina is present. Acute dilatation of one or both ventricles, again, in which stretching of the muscular fibres is obvious, frequently occurs without angina.

The importance, and even dominance, of this second unexplained factor becomes clear when the cases of advanced fatty degeneration are borne in mind, when the fibres must be incapable of producing anything like actual mechanical tension.

It has been assumed that the other element is neuralgic, and in a sense this is true, but not in the sense of predisposing neurotic tendency. It must be remembered that angina is much more common in the male, which is the least neurotic sex.

Mechanical stretching and neuralgic predisposition being put out of the question, there remains the fact that the existence of the patient is threatened at the moment of the attack by arrest of the heart's action, and were it not for the warning given by the pain and for the cessation of exertion enforced by it, the subject of the particular condition of the heart would die. We must, it seems to me, assume that angina is one of the defensive arrangements by which the adjustment of internal reactions to external conditions is secured.

Diagnosis.

Angina pectoris may be closely simulated by paroxysms of pain which are not symptomatic of disease of the heart of any kind, and are not attended with danger, and, as we usually have to depend on the account given by the patient, it is often a matter of great difficulty to distinguish between true angina and merely anginoid attacks. The difficulty will sometimes be aggravated by the fact that the patient has carefully read up the symptoms.

The age of the subject may be of assistance in determining the question. Angina is very rare before the age of forty-five, except in the case of aortic valvular disease or aneurysm or aortitis.

Sex, again, may often enable us to exclude angina without hesitation. It is extremely rare in women at any period of life in the absence of the conditions just enumerated, whereas so-called angina is a favourite complaint of neurotic ladies at all ages above thirty. What is described as "spasms" by tea-drinking female out-patients becomes angina among the educated and neurotic.

The appearance of the patient as ascertained from friends who have witnessed attacks may be valuable evidence. It does not necessarily follow that if he turns pale and has a look of alarm and suffering the paroxysms are those of true angina; but if his colour and expression show no change it will be evidence to the contrary.

The circumstances under which the early attacks come on are very significant. With rare exceptions the pain of angina is first experienced during exertion, and when it gradually increases in intensity with each successive attack and is provoked more and more readily, there can be little doubt as to its nature. If, on the other hand, the first paroxysms set in during repose, and particularly at a given interval after food, the inference is equally strong that they are pseudo-anginal in character and of gastric origin. Unless the history, onset, and nature of the paroxysms are quite characteristic, and confirmatory physical signs are present, we should only make a definite diagnosis of angina when all possible explanation of the pain can be excluded.

If we leave out of the consideration neurotic and hysterical attacks, which are usually easily recognizable, the cause of the spurious angina is nearly always some functional derangement of the stomach, and evidences of disturbance of the digestion, such as occasional attacks of vomiting, habitual flatulency with eructation, will often aid in establishing the distinctive diagnosis. In many cases dilatation of the stomach may be demonstrable by percussion and succussion.

A common combination is dilatation or distension of the stomach and high arterial tension. Together they give the nearest imitation of true angina, and if the heart be at all weak a fatal result is by no means impossible in elderly subjects. Such a result may be invited if the functional derangement of the stomach and liver are ignored, and digitalis or other cardiac tonic is given or the Schott treatment adopted, or if the paroxysms are simply treated by nitrite of amyl or nitro-glycerine.

Prognosis.

The **prognosis** of angina is beset with uncertainty. We can never tell when the next attack will come on, or whether it may not be the last. We are not, however, altogether without guidance, the elements of which will be an estimate of the relative predominance of the two chief factors in the production of the attack—whether inherent weakness of the heart wall on the one hand, or, on the other, obstruction in the circulation or other cause of embarrassment of the heart's action.

While the attacks only come on when provoked by exertion or excitement, or by flatulent indigestion (not, of course, taking the patient's word for the last-named cause), the hope may be entertained that by care in avoiding all known occasions they may be postponed indefinitely. If, further, there is habitual high tension in the pulse, this is at the same time evidence of obstruction in the arterioles and capillaries which may be capable of mitigation by treatment, and of some degree of vigour in the heart. So also will be accentuation of the aortic second sound, and still more any recognizable impulse or apex beat. The patient, of course, must not take exercise immediately after food, must never hurry or walk against a wind, and even on level ground must adapt his pace to his condition, and if compelled to go uphill must do so very gently and circumspectly.

Angina, again, in connection with aortic valvular disease, may run a very protracted course. It is when the pulse is soft and the heart is normal in dimensions, with imperceptible impulse and weak sounds—when, in fact, the results of careful examination are negative—that the greatest uncertainty and danger exist. The occurrence of unprovoked attacks and of nocturnal angina will emphasize this conclusion.

Treatment.

The primary significance of angina pectoris is, as has been said, that the heart is unequal to the task of propelling the blood. The heart is itself always in fault, but undue resistance in the vessels may play an important part in the production of the pain. The first consideration, therefore, when the treatment of angina is undertaken, will be whether there is arterio-capillary obstruction which can be removed. The pulse will be examined carefully and repeatedly at different periods of the day, before and after food, before and after the night's rest, before and after such exercise as the patient can take with impunity. If at any time the artery is distinctly full between the beats, virtual tension exists, *i.e.* there would be tension were there adequate *vis a tergo*, and it may be concluded that obstruction is present in the arterioles and capillaries which probably contributes to the embarrassment of the enfeebled heart, the removal of which may afford relief. Not uncommonly there will be found a well-marked senile pulse, with the arteries large, tortuous, and thickened, full between the beats, but compressible, the pulse wave being sudden and unsustained. Here the loss of elasticity and expansibility of the entire arterial tree will be a cause of difficulty to the heart. The aortic trunk and its main branches being atheromatous and refusing to dilate when the blood is propelled into them, the systole encounters the peripheral resistance at once just as if the vessels were a system of rigid, inelastic tubes. Nothing can be done to remedy the degeneration of the arteries, but it may be possible to lessen the resistance in the capillary net-work which has been a chief factor in the production of the atheromatous state, and which is now adding cardiac overstrain and angina to previous ill-effects.

In some cases of angina the pulse has all the characters of high tension without advanced disease of the vessels. This will usually be in gouty subjects, and we have angina which may justly be called "gouty."

Whenever high arterial tension can be traced in angina pectoris, there is an opening for treatment which may be palliative to a very important extent, and sometimes curative. The treatment will be such as has already been described in discussing high arterial tension. Colchicum may be given with the mercurial aperient in gouty angina, and also a mixture containing iodide and bicarbonate or citrate of potash, with gentian or some other vegetable bitter tonic, twice or three times a day. The iodide when well borne is often of remarkable service. In such cases the treatment may be pursued with confidence and with a certain degree of vigour. The diet must, of course, be strictly regulated, and heavy meals avoided.

When there is no conspicuous tension in the arteries, and their walls are in a state of degeneration, and when with this the walls of the heart are weak and probably fatty, while the same end is held in view and similar means are put in operation, great caution and watchfulness must be exercised. The bowels must be made to act daily, but mercurial aperients must be sparingly employed, an aloetic pill or liquorice powder, or some preparation of cascara, being given, if necessary, in the intervals.

When angina complicates disease of the aortic valves, it is difficult to say whether arterio-capillary resistance contributes in any way to its production, but if the pulse is good it will be well to give mild mercurial aperients on the assumption that such may be the case, though caution must be observed in their administration.

If the pulse in the intervals between the attacks of anginoid pain is small, short, easily compressible and destitute of tension, no good result is to be expected from eliminant treatment, and even small doses of mercury may depress the patient.

Prominence has been given to removal of arterio-capillary resistance by eliminant treatment, because when called for, it may yield more permanent relief than any other line of treatment; but arsenic and phosphorus may render very important service, and except in cases of markedly high arterial tension one or other of these should be given concurrently with eliminants. A particularly useful combination is arsenic with iodide of potassium and nux vomica. Phosphorus seems to have a specially favourable effect in angina associated with aortic regurgitation. Belladonna or cannabis indica may be useful adjuvants in some cases; quinine and nux vomica also are often of service. Nitro-glycerine and the heavier nitrites may also be required habitually, though it is best when possible to reserve their use for the anginoid attacks themselves.

Treatment of the Attacks.

There remains to be considered the treatment of the attacks themselves. Formerly brandy, various combinations of ether, nitrous ether, ammonia, lavender, and camphor, were the chief drugs resorted to. Inhalation of amyl nitrite and the administration of nitro-glycerine or of sodium nitrite have now almost entirely superseded these remedies.

Whatever the remedy, the patient should always carry it about with him, and have recourse to it as soon as the pain really sets in. The amyl nitrite is supplied in the convenient form of glass capsules containing five mins., enclosed in a silk bag, so that one of these can be broken in a handkerchief and the vapour inhaled. Some prefer to carry a small bottle of amyl nitrite about, which they can have recourse to when the attack threatens. Nitroglycerine, however, taken by the mouth, appears to be more generally useful, since, though the effect is scarcely as rapid as that of inhaled amyl nitrite, it lasts much longer. Tabloids of nitro-glycerine containing one min. of a one per cent. solution can be carried about, and one or two can be swallowed when necessary with very little loss of time.

In some cases nitro-glycerine has a better effect than amyl nitrite, though in rare instances nitro-glycerine appears to have no influence on the spasm to which nitrite of amyl at once gives relief. In two cases of the kind that I have seen, it has seemed to me extremely probable that the anginoid paroxysm had its origin in the right ventricle.

Occasionally when the nitrites fail we have to fall back on the old-fashioned remedies, especially when the heart failure is pronounced and the pulse tension is extremely low. Here it may do more good to help the heart by stimulants than to relieve it of work. While the paroxysms of angina are for the most part brief, the agony being such that it seems as if another moment must prove fatal, there are at times attacks of a protracted character. When the pain persists in spite of nitrites and stimulants, morphine and atropine should be administered hypodermically, and it is well to carry the needle into the substance of a muscle where the circulation is more active than in the subcutaneous cellular tissue. The initial dose should be small, but it may be necessary to employ morphine boldly. A turpentine stupe should also be applied over the region of the heart, or a mustard leaf or poultice.

It must be remembered also that the nitrites give rise to great frequency of the heart action which may be a source of distress. We should consequently employ them very cautiously when the angina is accompanied by a frequent pulse. The nitrites have been supposed to be heart tonics, but while their most prominent action is relaxation of the arterio-capillary net-work they also relax the cardiac muscular fibres.

As in so many other instances, the employment of nitroglycerine and the nitrites is not without its drawbacks. Patients often come to rely on the immunity from pain which the remedies confer and then presume upon it. Liberties are taken and imprudences are committed, so that not unfrequently sudden death is precipitated which might with care have been staved off for years. In placing the remedy in the patient's hands therefore, emphatic words of caution must be spoken and the danger must be pointed out.

The instructions to be given with regard to exercise are extremely important. After a very severe paroxysm, however provoked, rest in bed may be necessary for some days, and even for a period of two or three weeks, if the attack has been prolonged. The heart may be left extremely weak, its action slow or faltering and irregular, and the sounds scarcely audible, and sitting up or turning in bed may be attended with giddiness or faintness or pain in the region of the heart. When such conditions are present, time must be given to the heart to recover itself, and measures must be taken to relieve flatulence and constipation, which will probably be associated with the other symptoms.

Under ordinary circumstances, however, the rule usually applicable in heart disease holds good here. Whatever exercise the patient can take without provoking an attack at the time, or prostration afterwards, he will be the better for. While, however, the exercise should be as regular as possible, in no case is it more necessary to bear in mind the fact that the capacity for exertion varies from day to day, and that the sufferer from angina can do easily one day what would be impossible for him on another. This is one of the objections to the Œrtel methods of treatment. Some of the influences which affect him we can recognize, such as wind, or severe cold, or great heat, or weather which is felt by people in health to be oppressive, or a moisture-laden atmosphere; others arise out of internal conditions, flatulence, dyspepsia, constipation, functional derangements of the liver. The patient's feelings and inclinations have thus to be taken into account, but without allowing inertia or nervousness to have undue weight. There is great room here for judgment and tact and personal knowledge of the patient's disposition. Besides the caution necessary in cold and hot weather, the liability to anginoid attacks on walking soon after food must be borne in mind, and a period of rest after meals must be enjoined, especially after that particular meal which experience in the case under treatment has shown to be worst in this respect.

CHAPTER XXI.

FUNCTIONAL AFFECTIONS (SO-CALLED) OF THE HEART.

PAIN IN PRÆCORDIAL REGION—ITS CAUSES AND TREATMENT—PALPITATION : (1) PERSISTENT TACHYCARDIA ; (2) TEMPORARY INTERMITTENT ATTACKS OF PALPITATION—CAUSATION—TREATMENT—INTERMITTENT AND IRREGULAR ACTION OF THE HEART.

THE term "functional affections" is retained, not for any merit of its own but for want of a better. Under it must be discussed a variety of symptoms having the heart for their centre, but which cannot be assigned to any structural change. Taken all together they give rise to much actual suffering, and to far more nervous apprehension and fear of death than definite valvular and structural disease combined. So much is this the case that when patients come complaining of the heart, we are almost safe in concluding that the heart is disturbed by some cause outside itself and is not the seat of disease.

Pain is one of the symptoms which frequently gives rise to apprehension of heart disease. Leaving out of the question spurious angina, which has already been discussed, its most common seat is the region of the apex, but it may be felt over any part of the cardiac area, the left third space being next to the apex, the most frequent part in which pain is experienced. It is most commonly of a dull aching character, but may be sharp and stabbing or burning, and nervous women will exhaust all the epithets which can be applied to pain in their description of their sufferings. Tenderness on pressure very commonly accompanies the pain; it is superficial and is equally severe, whether the pressure is made over a rib or in an intercostal space; it is often particularly severe in the edge of the mamma when this extends into the tender area. The tenderness is quite extra-thoracic, and is felt when the heart is not even indirectly reached by the pressure. It is, therefore, a nervous hyperæsthesia, which may be a reflex from a distended or exhausted state of the heart, or may be subjective or symptomatic of some condition of the nervous system quite independent of any cardiac affection.

Another special seat of tenderness is over the second rib, about an inch from the edge of the sternum, where a branch of the cervical plexus crosses the rib. Pressure here not only causes pain, but may give rise to intolerable cardiac distress.

Causes of Cardiac Pain.

Pain in the region of the heart may be due to conditions of the heart itself, to direct pressure upon the heart by a dilated stomach or extreme distension of the abdomen, to reflex disturbance from some visceral derangement, or to nervous or emotional states.

Taking the last-named first, it is exemplified by the sharp pain in the heart, which may be induced by a sudden shock or fright, or by powerful emotion, and by the heartache of profound or protracted grief. But, without adequate emotional influence nothing is more common than cardiac pain as an expression of nervous depression.

Reflex pains are mostly of dyspeptic origin, but may be associated with uterine derangements. The pain caused by direct pressure of the diaphragm, carried upwards by a dilated stomach or distension of the colon or intestine generally, is accompanied by oppression of the breathing, and is usually felt at the base of the heart and is aggravated on lying down.

Pain due to overstrain of the heart is a diffuse ache over the cardiac area, generally accentuated in the region of the apex.

The treatment of cardiac pain will in its details vary with the cause. But in all cases it is most important to be able to convince the patient that there is no disease. While he has the idea in his mind that he is suffering from some serious heart affection, the concentration of his attention on the heart will be sufficient to renew the pain, and his apprehensions will interfere with the recovery of his nervous equilibrium.

Any derangement of the digestive organs or liver or uterus should be rectified by suitable diet and treatment, and, as a rule, tonics will be of service.

With the internal and general remedies the local application of belladonna as a liniment or plaster will be useful. The plaster is often more efficacious if it is applied so as to afford support or to exercise pressure on the painful part, and it is well, therefore, to apply it in strips.

Palpitation.

By palpitation is meant frequent and violent action of the heart, of which the subject is conscious; but patients will sometimes say they are suffering from palpitation when there is neither frequency nor violence recognizable in the beat of the heart or pulse by the observer, and, on the other hand, may be unconscious of extremely rapid action of the heart found on examination.

With palpitation there is usually uneasiness, sometimes pain, in the region of the heart, oppression of the respiration with frequent deep sighs and a sense of inability to fill the chest sufficiently. Often there is excitement and alarm and the patient feels giddy or faint; the face may be flushed or very pale.

When the heart is acting very rapidly, many of the beats may fail to reach the radial artery, so that the pulse becomes irregular. The reason probably is that the ventricle has not time to fill, and that consequently there is not sufficient blood propelled into the aorta to communicate a wave to the peripheral vessels. Very often the artery is small and full between the beats, there being a general excitement of the vascular system with spasm of the arterial walls.

On examination, during an access of palpitation, the heart may be felt to be beating violently, but when the rapidity of its action is extreme a faint vibration only may be communicated to the hand. On auscultation, the first sound may be loud and short, followed immediately by a weak second sound, or, in the case of extreme frequency, the first and second sounds may be almost identical in character and equidistant, resembling very closely those of the fœtal heart, and comparable to the puffing of a distant locomotive.

Attacks of palpitation may last a few minutes or many hours, and a special kind, for which the name tachycardia has been reserved, may go on for weeks or months and prove fatal. They almost always begin suddenly, sometimes after an apparent suspension of the heart's action, and they end suddenly or wear themselves out.

1. Tachycardia.

Tachycardia requires special notice, since the attacks which have come to be called by this name, evidently differ in kind as well as in degree from ordinary palpitation however severe. A frequency of 160, 180, 200, or even 240 may be maintained for many days. In a woman of about forty-five, the pulse frequency was never less than 200 for three weeks, and apparently this obtained during the short snatches of sleep. A retired naval officer consulted me from time to time for months, and when seen always had a pulse of 140. There was every reason to believe that this rate was maintained in the intervals between his visits, during which he went about as usual. Eventually dropsy set in and the patient died from heart failure.

Tachycardia is most common during middle age, but childhood and youth are not exempt from it. Grave's disease or exophthalmic goître, characterized by tachycardia, exophthalmos, and enlarged thyroid, will not be discussed here.

The main cause of tachycardia must reside in the nervous system, probably in the nervous apparatus of the heart itself, the cardiac ganglia, and nerves. No other adequate cause can be assigned. The naval officer of whom mention has been made, was gouty and had taken wine and spirits freely; the woman had had hard work as a lady's-maid, and the apparent exciting cause of the first attack was packing her mistress's box.

Treatment.

In tachycardia, digitalis and the cardiac tonics generally, or the carminatives which are useful in ordinary palpitation, appear to have little or no influence on the frequency of the heart action. Rest, mental as well as bodily, and simple diet must be insisted on, and any functional derangements of the abdominal or pelvic viscera should be corrected. The drugs which have seemed to exercise control over the heart have been bromides, in full doses—the sodium bromide being probably the safest and best—and belladonna or atropine, pushed to the limits of tolerance.

2. Temporary and Intermittent or Ordinary Palpitation.

Among the causes of what may be called ordinary palpitation as distinguished from tachycardia, the most important is a predisposition thereto on the part of the nerve centres governing the heart, which may be inborn or induced by modes of life or by the various circumstances which tend to lower the nervous tone or to promote nervous excitability. Palpitation is much more common in women than men, partly in virtue of the greater inherent susceptibility of the female nervous system, partly from the more emotional life of women, their greater confinement to the house, and their less vigorous exercise; but child-bearing and over-lactation are also in themselves serious predisposing causes. In men a sedentary mode of life, exciting occupations, dissipation, and excesses of all kinds, over-indulgence in tobacco, bring about a liability to palpitation.

Among the exciting causes are sudden violent impressions on the nervous system of any kind—fright, an unexpected noise, a startling incident taking place before the eyes, a powerful emotion; these will set any one's heart beating, but in a strong and healthy person the effect is of very brief duration; where the predisposition to palpitation exists they may initiate an uncontrollable attack. A similar statement applies to exertion—a brisk walk uphill causes the heart to act rapidly and powerfully under normal circumstances; in a predisposed individual the action may be exaggerated and protracted, so as to constitute an attack of palpitation.

But the characteristic palpitation of the heart starts suddenly without obvious exciting cause, while the patient is sitting quietly at work or reading, or during sleep, when the patient may wake up from a frightful dream which appears to have brought on the attack.

There is, however, as a rule, some internal exciting cause, which is most commonly some gastric derangement attended or not with flatulent distension of the stomach or bowels. Other forms of peripheral irritation may act as exciting causes, such as uterine affections.

To arrest an attack of palpitation it is sometimes only necessary to take a dozen deliberate deep breaths, and it is always well to try this expedient before resorting to drugs. The remedies of most general service are combination of alkaline and carminative stimulants; bicarbonate of soda with ammonia and camphor or peppermint water is often sufficient, but compound tincture of chloroform, ether, valerian, lavender, ginger may be added or substituted; in some cases bromides are required. Undue acidity of the gastric contents is corrected, flatulence is expelled, and possibly the stimulation of the pneumogastric fibres of the mucous membrane of the stomach may have some inhibiting influence on the heart. Digitalis appears to have little or no effect, but belladonna may be useful, especially in combination with bromide of ammonium or sodium.

For the prevention of palpitation the tone of the nervous system must be raised by the usual hygienic and medicinal means, namely, change to the seaside, or, better, to mountain health resorts, exercise, fresh air, early hours, simple wholesome food, avoidance of excitement of all kinds. Tonics and remedies for the functional derangements of liver, stomach, or other organs should also be given, when necessary. The emplast. belladonna over the region of the heart appears to have some influence in preventing or moderating the attack.

A complaint made more frequently by women than men is of a feeling that the heart is stopping; sometimes it is a faltering for a few beats, sometimes a sense of total arrest. The pulse need not be affected at all at the time, but it may be unduly frequent, or very slow, or intermitting.

The causes and treatment of this affection are so similar to those of palpitation that no special consideration of them is necessary.

Intermittency of the Pulse and of the Action of the Heart.

By an intermittent pulse is meant a pulse in which a beat is missing from time to time, while in the intervals it is perfectly regular. The intermission may occur at regular and definite periods every four, six, or more beats up to twenty, or the number of intervening pulsations may vary.

It is not characteristic of any form of heart disease and is rarely indicative of organic disease. An intermittent pulse may be constant and habitual, and the intermission is then more likely to occur at definite intervals; it may be occasional only, and may be attributable to some disturbing reflex cause, of which flatulent dyspepsia is the most common. In some cases the pulse is intermittent after each meal; or in others tea, coffee, or tobacco may be the special cause of the intermission. It is common also in chronic gout, and may be among the signs of fatty degeneration of the heart. In case of doubt the patient should be made to walk briskly for a minute or two, when, if the heart is really weak and degenerated, the pulse will falter, whereas, if the heart is healthy, the intermission will usually disappear. Intermittency of pulse may also be associated with nervous debility and hypochondriasis, the pulse becoming normal again when the patient regains good health. On examining the heart it is usually found that the cause of the intermission is not the actual omission of a heart beat, but the occurrence of a hurried and imperfect contraction which rapidly follows the last of the series of normal beats, and does not transmit a pulse wave to the wrist. The imperfect beat may sometimes be felt on palpation; usually only the first sound is heard on auscultation at the apex unaccompanied by a second sound. The heart beat which follows the intermission is usually more powerful than normal. While it is the rule that there is this feeble interposed heart beat, instances occur where it cannot be heard or felt, and in which the heart appears to remain quite passive.

The patient may or may not be conscious of the intermittent action of the heart. He is more likely to be aware of it when it is symptomatic of some functional derangement than when it is habitual; he may be conscious of a vague sense of discomfort or of an unpleasant sinking sensation in the cardiac region during the intermission, or he may feel the bump of the stronger beat which usually follows the feeble and imperfect or dropped beat. The occurrence of the intermission is difficult to explain, and we are compelled to fall back on nervous influence as the agent in its causation. When the intermittency is constant and habitual it appears to have no significance in relation either to the heart, nervous system, or vital power generally, and it may be met with in men who enjoy vigorous health and live to a good old age. It is unnecessary to say that in such cases no treatment is required. When the intermittent action of the heart is only occasional it is usually traceable to some exciting cause, such as tea, coffee, or tobacco, or to dyspepsia and flatulence, and suitable treatment for removal of the cause should then be adopted.

IRREGULARITY OF THE HEART'S ACTION.

Irregularity of the heart's action, like intermittency, may be habitual or occasional. Habitual irregularity, whilst it is commonly present in mitral incompetence and in cases of cardiac dilatation of any severity, may occasionally be met with in individuals in whom there is no evidence of any heart disease, and who enjoy good health and live to old age. In one instance I have known the heart's action, without any apparent cause, to be markedly irregular during a period of at least twenty years to my knowledge in a gentleman who lived to the age of seventy and enjoyed vigorous health.

Temporary irregularity of the cardiac action is much more common, and may be occasioned by tobacco or strong tea, by mechanical embarrassment of the heart by a distended stomach or colon, or by various reflex causes, such as gastric and liver derangements, or emotional disturbances of various kinds.

Irregularity of the cardiac action is more serious than intermission, and steps should at once be taken to remove, if possible, the exciting cause. When there is mechanical embarrassment of the heart by a distended and dilated stomach the distress at times may be severe, and there is danger of a sudden syncopal attack in elderly people in whom the heart has undergone degenerative changes.

APPENDIX.

Note on the Preparation of the Baths, and on the Movements practised in the Schott Treatment of Heart Disease.

For a weak bath, with which the treatment is usually commenced, about 1 lb. of common salt and 1½ ozs. of chloride of calcium should be added for every 10 gallons of water used. The temperature of the first bath should be 92° to 95° Fahr., and its duration about six minutes. The baths may be given every other day, or every day for four or five days, when a day's rest should be allowed. They should gradually be increased in strength, till the maximum of 3 lbs. of salt and 4½ ozs. of chloride of calcium for every 10 gallons of water is attained. The temperature of each successive bath should be lowered and its duration prolonged, according as the patient bears it, till the temperature is about 85° Fahr. and the duration about twenty minutes. After each bath a period of rest in the recumbent position should be enjoined.

For the preparation of the effervescing baths which should succeed the plain saline baths, bicarbonate of soda and hydrochloric acid may be used, about 2 ozs. of the former and 3 ozs. of the latter being added for every 10 gallons of water for the weakest effervescing baths. The bicarbonate of soda should be dissolved in the brine bath of full strength, and the acid, previously diluted, should be gradually added and well mixed with the water just before use. The amount of each ingredient may be gradually increased daily, till the maximum of 8 ozs. of bicarbonate of soda and 12 ozs. of hydrochloric acid for every 10 gallons is attained. This, however, is a somewhat inconvenient method, as the fumes of the acid may be irritating and its strength varies; a German chemist has prepared tablets in which the acid is incorporated in a convenient and portable form, which are known as "Sandow's Tablets," and from these, together with the alkaline powder also supplied, effervescing baths of any strength can be conveniently and safely prepared.

The duration of the course of treatment is about a month or five weeks, and the effervescing baths may usually be commenced after the first fortnight, or sooner in some cases, though in more severe cases they may have to be deferred for a longer period or not employed at all.

The Movements or Exercises.

The exercises consist of a series of simple movements of each limb and of the trunk in turn made against slight resistance afforded by a trained attendant, so that each muscle in the body, as far as this is possible, is in turn brought into action.

The movements should be made slowly and systematically, no movement being repeated twice in succession, and a short interval should be interposed between each one. They should be stopped for a time if the patient experiences any distress or discomfort in respiration, and should only be resumed when he has rested sufficiently. The patient should be told to breathe regularly and deeply, as far as possible. The movements are as follows :—

1. The arms are extended in front of the body at the level of the shoulder with the palms of the hands touching. The two arms are then moved slowly outwards till they are in a line with each other; they are then brought back to their original position.

2. The arms being extended at the side of the body with the palms turned outwards, they are abducted and raised till the hands meet above the head: they are then brought back again in the same way.

3. Each arm in turn, extended by the side of the body with the palm turned forwards, is flexed at the elbow till the fingers touch the elbow, and is then extended again.

4. The arms being by the side with the palms of the hands turned inwards, they are rotated forwards and upwards at the shoulder-joint till they are vertically extended above the head, parallel to each other : they are then depressed in the same way.

5. The hands being clinched and turned outwards and the arms extended, the forearm is flexed till the fingers touch the shoulder : it is then again extended.

6. The arm is rotated at the shoulder-joint forwards and

upwards and then backwards and downwards till a complete revolution has been performed.

7. The arms being by the side and the palms turned inwards, they are moved upwards and backwards as far as possible, and are then brought back again.

8. The trunk is flexed on the hips, the knees being kept stiff, and is then extended again.

9. The trunk is rotated laterally first to the right and then to the left.

10. The trunk is flexed laterally first on one side and then on the other.

11. Each leg is flexed in turn at the hip-joint, the knee being bent.

12. The knee being kept straight, each leg in turn is raised as high as possible in front of the body, and then in the same way behind.

13. Each leg in turn is then abducted as far as possible, the knees being kept straight.

14. Each knee in turn is flexed, the patient standing on each leg alternately and supporting himself with a chair.

15. Finally, flexion and extension of the wrists and of the ankles may be practised.

The assistant stands opposite the patient, and, by placing his hands on the limb which is being exercised, makes gentle resistance to each movement as it is performed.

This descriptive note is appended, not with the object of giving undue importance to this form of treatment for heart disease, but in the hope that it may be of service to those who have not had an opportunity of seeing it carried out, so that if they wish they can try it for themselves.

The mere fact that a patient has heart disease should not be the signal for its immediate employment. When compensation is perfect no special treatment is required, and in a large proportion of cases of real valvular and structural disease, the value and efficacy of the Schott treatment are doubtful, and equally good or better results can be obtained by other methods of treatment. The best results will be obtained in cases of functional or imaginary heart disease in neurotic individuals.

Too much importance is attached by advocates of the Schott treatment to the percussing out of the area of cardiac dulness and to the diminution it is said to undergo after each bath, more especially when the so-called auscultatory method is employed. This method lends itself very much to the imagination, and is absolutely untrustworthy. A shifting inwards of the apex beat is of importance, but it is probable that the diminution of percussion dulness is due, not so much to fluctuations in the size of the heart, as to encroachment by the lungs on the cardiac area due to deeper respirations taken while the patient is in the bath. It does not therefore follow that, because the area of cardiac dulness diminishes after a bath, the heart was previously dilated. Accurate delineation of the outline of the heart by percussion is in many cases impossible, and even in the mere percussing out of the area of superficial cardiac dulness there may be many sources of error, so that a diagnosis of cardiac dilatation should not be made from the evidence of percussion alone. The practice of placing in the hands of the patient elaborate diagrams to illustrate the supposed diminution in the size of the heart after a bath is to be deprecated.

INDEX.

A

Accent at beginning of a murmur, 32
Accentuation of aortic second sound, 34
—— of pulmonic second sound, 33, 169, 191
Adherent pericardium, 220
Ætiology of valvular lesions, 60
Age, as affecting hypertrophy, 39
——, —— prognosis, 44, 68, 78, 151, 201
——, —— treatment, 101
——, old, mitral incompetence of, 179
Alcohol, 98, 106, 182, 269, 285
Anæmia, 65, 82, 99, 148, 176, 242, 256, 268, 285
Angina pectoris, 296
—— —— in aortic incompetence, 48, 160
——, pseudo-, 307, 316
—— vasomotoria, 305
Aortic area, sounds over, 24
—— incompetence, 137
—— ——, changes in the heart in, 42
—— ——, pulse in, 35
—— ——, symptoms in, 48
—— ——, use of digitalis in, 117
—— —— with stenosis, 131, 156
—— stenosis, 124
—— ——, hypertrophy in, 41
—— ——, pulse in, 35
Aortitis, acute, 127, 302
Aperients, 108, 263, 295
Apex beat, displacement of, 18
——, murmurs at, 23
——, sounds at, 21
Apoplectiform seizures, 288
Apoplexy, pulmonary, 50, 87, 202
Arterial tension, effects of high, 34, 40, 62, 80, 149, 173, 182, 237, 242, 263, 305
Asystole, 110
Atheroma, 61, 146
Auricle, left, 13, 44
——, right, 12
Auscultation, 21

B

Balfour, 65
Baths, 92, and appendix
Bed, rest in, 89, 103, 157, 267, 313
Bradbury, 111
Breath, shortness of, 48, 49. 217, 243, 256
Bronchitis, 33, 182, 203, 217

C

Caffein, 113
Capillary pulsation, 141
Cardiac dulness, deep, 13, 252
—— ——, superficial, 15, 20
—— pain, 316
—— tonics, 113
Childhood, rheumatism in, 78, 89, 158
Climate, 96, 181
Clinical examination of heart, 15
Cœruleus, morbus, 215
Compensation, 46, 91, 95, 157, 170, 201, 225, 235
Congenital affections of heart, 213
Coupled heart-beats, 123, 206
Cyanosis, 49, 215, 257

D

Danger, relative, of valvular lesions, 68
Death, sudden, 69, 74, 150, 160, 279, 286, 307
Degeneration, fatty, of heart, 282
Degenerative changes, 61, 134, 146, 149, 156, 237, 278
Diet, 97, 106, 266
—— in Œrtel treatment, 91
Digitalis, 112
—— in aortic disease, 117, 136, 161
—— in mitral disease, 118, 183, 206
Dilatation of heart, 38, 238
—— —— in aortic incompetence, 43, 146, 240

INDEX.

Dilatation of heart in mitral incompetence, 45, 170
—— of right ventricle, 272
Dilators, vascular. *Vide* Vaso-dilators
Dropsy, 50
Duckworth, 186

E

Embolism, 197, 204
Endocarditis, acute, 60, 72
——, chronic, 61, 73
Epileptiform attacks, 288
Exercise, 90, 100, 181, 204, 270, 313
Exercises, Schott, 93, and appendix

F

Fatty infiltration of heart, 92, 282
—— degeneration of heart, 282
Frequency, relative, of valvular lesions, 27
Functional affections of heart, 95, 315

G

Gairdner, 185

H

Hæmic murmurs, 126, 175
Hæmoptysis, 50, 204
Heart, cavities of, 11
——, dilatation of, 38, 238
——, fatty degeneration of, 282
——, —— infiltration of, 92, 282
——, hypertrophy of, 38, 230
——, relations of, 13
Heredity, 79
Hypertrophy, 38, 230
—— in aortic incompetence, 42, 147
—— in aortic stenosis, 42, 130
—— of right ventricle, 272

I

Inspection, examination by, 15
Intermittent action of heart, 34, 322
Irregular pulse. *Vide* Pulse

J

Jugular veins, distension and pulsation of, 16, 47, 208, 209, 258

L

Leeches, use of, 110, 183, 206, 262
Liver, enlargement of, 19, 47, 108, 172, 197, 208, 258
——, pulsation of, 20, 47, 108, 197, 208

M

McAlister, 64, 175
Mackenzie, 17, 209
Mitral area, sounds over, 21
—— incompetence, 162
—— ——, changes of heart in, 44
—— ——, digitalis in, 118
—— ——, from stretching of orifice, 63, 76, 174
—— ——, pulse in, 35
—— —— symptoms in, 49
—— stenosis, 185
—— ——, changes of heart in, 45
—— ——, digitalis in, 122
—— ——, pulse in, 35
—— ——, stages of, 191
Morphia, 160, 312
Murmurs, cardiac, 29
——, diastolic aortic, 23, 24, 31, 138, 210
——, hæmic, 126, 175
——, presystolic, 23, 139, 191
——, pulmonic, 210
——, retarded systolic, 32, 168, 178
——, systolic, apical, 23, 162, 241, 253
——, ——, basic, 23, 124, 128, 210
——, ——, late in life, 74, 128, 179

N

Nauheim, 92
Nitrites. *Vide* Vaso-dilators
Nodules rheumatic, 89, 158

O

Œdema, 59
Œrtel, 91

P

Pain, præcordial, 316
Palpation, 17, 232, 251, 316
Palpitation, 256, 317
Peacock, 214, 215, 218
Percussion, 20
Pericardium, adherent, 79, 121, 220 279, 301

INDEX.

Presystolic murmur. *Vide* Murmurs
—— thrill, 17, 191
Prognosis, 67
—— in angina pectoris, 307
—— in aortic incompetence, 152
—— —— —— with stenosis, 156
—— —— stenosis, 133
—— in dilatation of heart, 258, 267
—— in endocarditis, 78, 83
—— in fatty degeneration of heart, 291
—— in mitral incompetence, 172
—— —— stenosis, 200
Progressive lesions, 61, 73, 156, 201, 289
Pulmonic area, sounds over, 25
—— incompetence, 142, 210
—— stenosis, 210, 213, 216
Pulsation, violent arterial, 140
Pulse, 35
——, collapsing, 35, 37, 131, 144, 145, 152, 157
——, Corrigan's, 141
——, delay of, 142, 150
—— in aortic incompetence, 35, 141
—— —— stenosis, 35, 131
—— in mitral incompetence, 35, 165
—— —— stenosis, 35, 188
——, intermittent, 34, 322
——, irregular, 34, 323
——, ——, in aortic incompetence, 143
——, ——, in cardiac dilatation, 250, 258
——, ——, in mitral incompetence, 37, 121
——, ——, —— stenosis, 189, 199, 205
Pulsus bisferiens, 142, 155
Purgatives, 108, 262

R

Residence, choice of, 96
Rheumatism, 60, 78, 89, 95, 158, 203
Rupture of heart, 286
—— of valve, 63, 74, 150

S

Schott treatment, 92, and appendix
Sex as affecting prognosis, 79
Sleeplessness, 49, 50, 159, 258
Sounds of heart, 21
—— ——, interval between, 22, 254, 318
—— ——, modifications of, 32, 128, 150, 168, 191, 233, 252, 318
—— ——, reduplication of, 22, 24, 34

Stomach, distension of, 98, 282, 286, 295, 313, 316, 321
Strophanthus, 113
Sudden death. *Vide* Death
Symptoms in angina pectoris, 296
—— in aortic disease, 48, 132, 150
—— in dilatation of heart, 255
—— in fatty degeneration, 286
—— in mitral disease, 49, 172, 196
—— in rupture of valve, 74
Syncopal attacks, 74, 160, 161, 287
Syphilis, 135, 149, 285
Systolic murmurs. *Vide* Murmurs

T

Tachycardia, 318
Temperature, changes of, 95, 300, 313
Treatment, 88
—— after endocarditis, 89
—— in angina pectoris, 160, 308
—— in aortic incompetence, 157
—— —— stenosis, 135
—— in dilatation of heart, 261, 269
—— in fatty degeneration, 292
—— in functional affections, 317, 319, 321
—— in mitral incompetence, 180
—— —— stenosis, 203
——, Schott, 92, and appendix
——, Œrtel, 91
—— for venous obstruction, 108, 183
Tricuspid area, 24
—— incompetence, 207
—— stenosis, 209

V

Vaso-dilators, 111, 136, 160, 206, 258, 311
Venesection, 109, 161, 183, 205, 261
Venous stasis, 49, 108, 205, 257
Ventricle, left, 12
——, ——, dilatation of, 239
——, ——, dilatation and hypertrophy of, 42, 147, 170, 240
——, ——, hypertrophy of, 230, 130
——, right, 12
——, ——, dilatation of, 272
——, ——, dilatation and hypertrophy of, 45, 170, 187, 214, 222, 272
Vomiting, 49, 161, 259

W

Walshe, 27, 51, 68, 301
Water-hammer pulse, 141

LONDON:
PRINTED BY WILLIAM CLOWES AND SONS, LIMITED,
STAMFORD STREET AND CHARING CROSS.